Field Guide to

Interferometric Optical Testing

Eric P. Goodwin
James C. Wyant

SPIE Field Guides
Volume FG10

John E. Greivenkamp, Series Editor

Bellingham, Washington USA

Library of Congress Cataloging-in-Publication Data

Goodwin, Eric P.
Field guide to interferometric optical testing / Eric P. Goodwin & James C. Wyant.
p. cm. -- (The field guide series ; 10)
Includes bibliographical references and index.
ISBN 0-8194-6510-0
1. Optical instruments--Testing. 2. Interferometry. I. Wyant, James C. II. Title.

TS514.G66 2004
535'.470287--dc22

2006024169

Published by

SPIE—The International Society for Optical Engineering
P.O. Box 10
Bellingham, Washington 98227-0010 USA
Phone: +1 360 676 3290
Fax: +1 360 647 1445
Email: spie@spie.org
Web: http://spie.org

Printed in the United States of America.

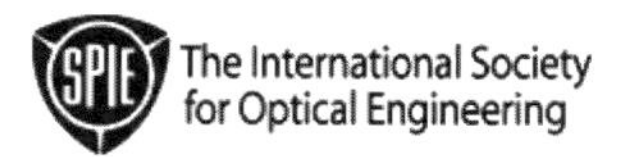

Introduction to the Series

Welcome to the *SPIE Field Guides*—a series of publications written directly for the practicing engineer or scientist. Many textbooks and professional reference books cover optical principles and techniques in depth. The aim of the *SPIE Field Guides* is to distill this information, providing readers with a handy desk or briefcase reference that provides basic, essential information about optical principles, techniques, or phenomena, including definitions and descriptions, key equations, illustrations, application examples, design considerations, and additional resources. A significant effort will be made to provide a consistent notation and style between volumes in the series.

Each *SPIE Field Guide* addresses a major field of optical science and technology. The concept of these *Field Guides* is a format-intensive presentation based on figures and equations supplemented by concise explanations. In most cases, this modular approach places a single topic on a page, and provides full coverage of that topic on that page. Highlights, insights, and rules of thumb are displayed in sidebars to the main text. The appendices at the end of each *Field Guide* provide additional information such as related material outside the main scope of the volume, key mathematical relationships, and alternative methods. While complete in their coverage, the concise presentation may not be appropriate for those new to the field.

The *SPIE Field Guides* are intended to be living documents. The modular page-based presentation format allows them to be easily updated and expanded. We are interested in your suggestions for new *Field Guide* topics as well as what material should be added to an individual volume to make these *Field Guides* more useful to you. Please contact us at fieldguides@SPIE.org.

John E. Greivenkamp, *Series Editor*
Optical Sciences Center
The University of Arizona

The Field Guide Series

Keep information at your fingertips with all of the titles in the Field Guide Series:

Field Guide to Geometrical Optics, John E. Greivenkamp (FG01)

Field Guide to Atmospheric Optics, Larry C. Andrews (FG02)

Field Guide to Adaptive Optics, Robert K. Tyson and Benjamin W. Frazier (FG03)

Field Guide to Visual and Ophthalmic Optics, Jim Schwiegerling (FG04)

Field Guide to Polarization, Edward Collett (FG05)

Field Guide to Optical Lithography, Chris A. Mack (FG06)

Field Guide to Optical Thin Films, Ronald R. Willey (FG07)

Field Guide to Spectroscopy, David W. Ball (FG08)

Field Guide to Infrared Systems, Arnold Daniels (FG09)

Field Guide to Interferometric Optical Testing, Eric P. Goodwin and James C. Wyant (FG10)

The material covered in the *Field Guide to Interferometric Optical Testing* is derived from a course taught by Dr. Wyant at the College of Optical Sciences at the University of Arizona. The material has evolved over the years as the underlying technologies and techniques have changed. This text is meant as a reference of interferometric principles and methods for the practicing engineer.

Eric Goodwin dedicates this *Field Guide* to his wife, Sam, and their daughter, Ryan.

James Wyant dedicates this *Field Guide* to the memory of Louise.

Eric P. Goodwin and James C. Wyant
College of Optical Sciences
University of Arizona

Table of Contents

Table of Contents

Table of Contents

Glossary

Frequently used variables and symbols:

a	Average phase shift between frames
A	Amplitude
A_n	Aspheric surface coefficients
b	Number of bits for quantization error
B	Obscuration ratio
c	Speed of light
C	Moiré fringe spacing
C	Curvature
d	Distance, displacement
D	Diameter
D_{HS}	Diameter of Hindle Sphere
f	Focal length
f	Spatial frequency
$f/\#$	F-number
F	Focal point
F	Coefficient of finesse
$g[\theta']$	Zernike angular component
G	G-factor
h	Height
H	Normalized field height
i	Step number, frame number
I	Irradiance
L_c	Coherence length
m	Diffraction order or fringe order
m	Fresnel zone plate zone number
m	Transverse or lateral magnification
n	Index of refraction
n_e	Extraordinary index, uniaxial crystal
n_o	Ordinary index, uniaxial crystal
N	Number of algorithm steps
N	Integer number of 2π
NA	Numerical aperture
OPD	Optical path difference
OPL	Optical path length
p	p-polarization state
r	Non-normalized radial coordinate
r_m	Radius of mth bright fringe
r_p	Pupil radius

Glossary (cont'd)

R	Radius of curvature
R/T	Reflection/transmission ratio
R_s	Radial shear coefficient
s	s-polarization state
$s(r)$	Sag as function of part radius
S	Fringe spacing
SNR	Signal to noise ratio
t	Thickness
t_c	Coherence time
T	Lateral translation
v	Speed of light in medium or velocity
V	Visibility
V_{sc}	Visibility factor due to spatial coherence
W_{ijk}	Wavefront aberration coefficients
$W(x, y)$	Wavefront as function of spatial position
x	Spatial coordinate
x_p	Pupil coordinate
x_s	Pixel spacing
x_w	Pixel width
y	Spatial coordinate
y_p	Pupil coordinate
z	Object distance, axial position
z'	Image distance (lens)
Z	Zernike polynomial coefficients
α	Angle between two polarization states
α	Moiré angle, wedge angle
β	Tilt
$\delta\beta$	Tilt difference
Γ	Fringe contrast
$\delta(x, y)$	Grating errors, function of position
Δ	Fringe displacement
Δ	Integrated phase change
ε	Linear phase shift error
ε	Angle error for 90-degree prism
ε_z	Axial distance from paraxial focus
η	Diffraction efficiency
θ	Angle, shear angle, tilt orientation
θ'	Angle, Zernike polynomial set

Glossary (cont'd)

θ_d	Diffraction angle
θ_i	Incident angle
κ	Conic constant
λ	Wavelength
λ_c	Center wavelength
λ_{eq}	Equivalent wavelength
Λ	Diffraction grating or moiré grating period
υ	Frequency
$\Delta\upsilon$	Frequency difference
ξ_c	Cutoff frequency
$\xi_{c,sa}$	Cutoff frequency for a sparse array detector
ξ_{Ny}	Nyquist frequency
ρ	Reflectance (ratio of reflected irradiance)
ρ	Normalized pupil radius ($0 < \rho < 1$)
σ	RMS wavefront error
σ^2	Wavefront variance
$\sigma_{\phi,i}$	Standard deviation, irradiance fluctuations
$\sigma_{\phi,q}$	Standard deviation, quantization phase error
ϕ	Phase
$\phi(t)$	Phase shift as a function of time
Ω	Rotation rate
Ω	Solid angle

Two-Beam Interference Equation

Interferometric optical testing is based on the phenomena of interference. **Two-beam interference** is the superposition of two waves, such as the disturbance of the surface of a pond by a small rock encountering a similar pattern from a second rock. When two wave crests reach the same point simultaneously, the wave height is the sum of the two individual waves. Conversely, a wave trough and a wave crest reaching a point simultaneously will cancel each other out. Water, sound, and light waves all exhibit interference, but for the purpose of optical testing, the focus will be the interference of light and its applications.

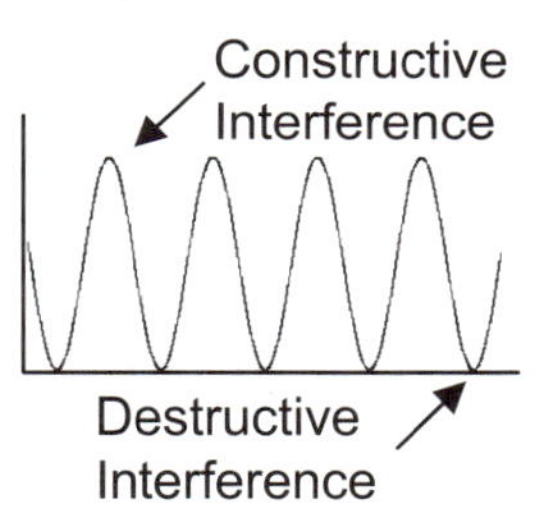

A light wave can be described by its **frequency**, **amplitude**, and **phase**, and the resulting interference pattern between two waves depends on these properties, among others. The **two-beam interference equation** for monochromatic waves is:

$$I(x,y) = I_1 + I_2 + 2\sqrt{I_1 I_2}\cos(\phi_1 - \phi_2)$$

$$I(x,y) = A_1^2 + A_2^2 + 2A_1A_2\cos(\phi_1 - \phi_2)$$

- I is the **irradiance**. Detectors respond to irradiance, which is the electric field **amplitude**, A, squared:

$$I = A^2$$

- ϕ is the **phase** of the wave in radians:

$$0 \le \phi \le 2\pi$$

- $\phi_1 - \phi_2 = \Delta\phi$ is the phase difference between the test and reference beams

Basic Concepts and Definitions

- λ is the wavelength of light. Visible light extends from 400 to 700 nanometers

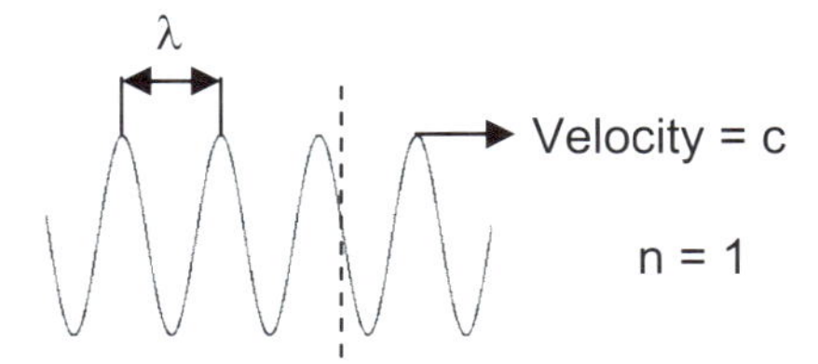

Wave Crests / Second = υ

- c is the speed of light in a vacuum ($n = 1$):

$$c = 2.99792 \times 10^8 \frac{\text{m}}{\text{s}}$$

- υ is the optical frequency of the light:

$$\upsilon = \frac{c}{\lambda}; \quad \upsilon = 5.45 \times 10^{14} \text{ Hz for } \lambda = 550 \text{ nm}$$

- n is the **index of refraction** of the medium. n is a function of λ:

$$n(\lambda) = \frac{c}{\text{v}} = \frac{\text{Speed of Light in Vacuum}}{\text{Speed of Light in Medium}}$$

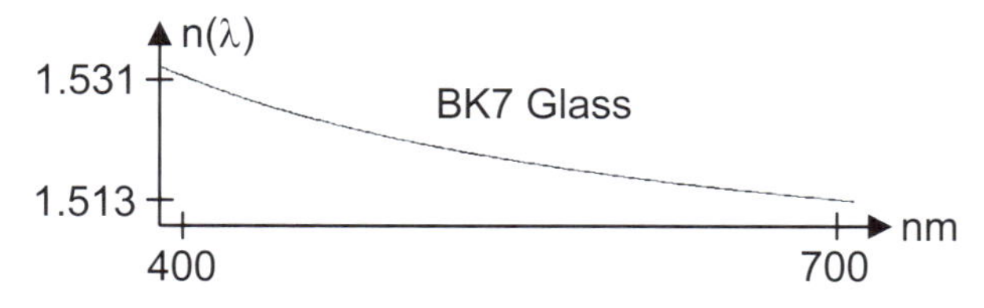

- OPL is the **optical path length** and is proportional to the time light takes to travel from a to b:

ds
a
n(s)
b

$$OPL = \int_a^b n(s)ds; \quad OPL = nt$$

- t is the physical thickness of the medium
- OPD is the **optical path difference** between two beams:

$$OPD = OPL_1 - OPL_2; \quad (\phi_1 - \phi_2) = \frac{2\pi}{\lambda} OPD$$

Conditions for Obtaining Fringes

In order for **interference fringes** to be observed between two beams, several conditions must be met. The light in one beam must be both **temporally** and **spatially coherent** with the other beam in the region where interference fringes are to be observed. In addition, the **polarization** properties of the two beams must be compatible. Finally, the relative irradiances of the two beams must be close in magnitude.

Temporal coherence is inversely proportional to the **spectral bandwidth** of the light source of the two beams. Temporal coherence goes as the **Fourier transform** of the spectral distribution of the source. For example, a **laser** is often modeled as a purely monochromatic source; that is, its spectral bandwidth is zero. The Fourier transform of a zero bandwidth source is a constant, so the temporal coherence of a purely monochromatic source is infinite. Infinite temporal coherence means that light in the one beam can be delayed relative to the second by any amount of OPL (time) and the two beams will still interfere. The temporal coherence of a source is usually given by the **coherence length** (L_c) or the **coherence time** (t_c):

$$L_c = \frac{\lambda_c^2}{\Delta\lambda} \qquad t_c = \frac{L_c}{c}$$

Source	λ_c (nm)	$\Delta\lambda$	L_c
HeNe Laser	632.8	<0.04 pm	>10 m
Hg Lamp	546	~0.1 nm	~3 mm
SLD	680	12 nm	38 μm
LED	660	25 nm	17 μm
Light Bulb	550	~300 nm	~1 μm

The center wavelength is λ_c and $\Delta\lambda$ is the spectral bandwidth, measured at the FWHM. A source can produce fringes as long as the OPD between the two beams is less than the coherence length. If the magnitude of the OPD between the two beams is greater than the coherence length, fringes will not be observed. As the OPD goes to zero, fringe **visibility** reaches a maximum.

Visibility

Visibility ranges from 0 to 1, where fringes with $V > 0.2$ are usually discernible. The two-beam interference equation should be modified by a temporal visibility function, which depends on the source used and is a function of the OPD. For a laser, $V(OPD)$ goes to 1 and the original two-beam interference equation remains:

$$V = \frac{I_{\max} - I_{\min}}{I_{\max} + I_{\min}}; \quad I = I_1 + I_2 + 2 \cdot V(OPD) \cdot \sqrt{I_1 I_2} \cos(\Delta\phi)$$

Fringe visibility variations due to temporal coherence

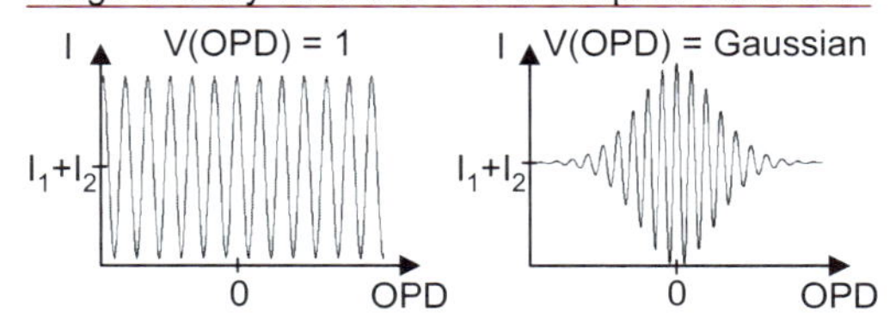

Fringe visibility degradation due to temporal coherence can be improved by decreasing the OPD between the two beams or by spectrally filtering the source.

Spatial Coherence

A second requirement for the observation of interference fringes is that the two beams are spatially coherent. If the source is truly a point source, the two beams are identical except for the OPD. The model of an ideal monochromatic point source gives two beams in which any location in the first beam will interfere with any location in the second beam. Point sources do not exist and real beams have less than ideal spatial coherence. The visibility of the interference fringes for two beams is inversely proportional to the spatial extent of the light source.

Fringe visibility variations due to spatial coherence

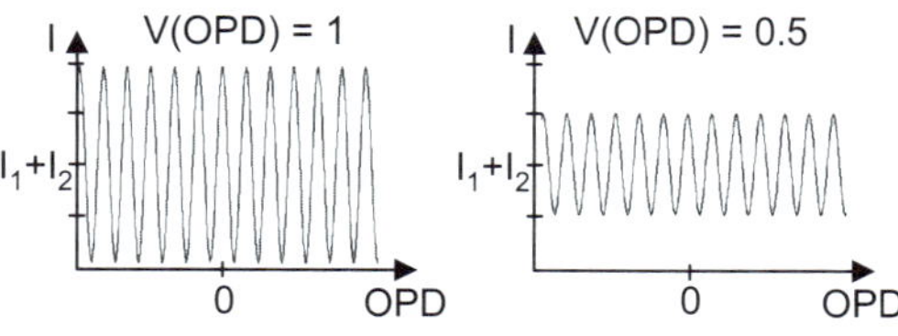

Spatial Coherence

Unlike fringe visibility for temporal coherence, fringe visibility degradation due to spatial coherence effects is not a function of OPD. This degradation can be represented in the two-beam interference equation by a constant V_{sc} $(0 \leq V_{sc} \leq 1)$ multiplied by the cosine term:

$$I = I_1 + I_2 + 2 \cdot V_{sc} \cdot \sqrt{I_1 I_2}\cos(\Delta\phi)$$

The spatial coherence of a source can be improved by spatially filtering the light source. One way to spatially filter a light source is to use a lens to couple the light into a single mode fiber. Light emerging from the other end of the fiber will fill the numerical aperture of the fiber but will be originating from the central core of the fiber, which is a good approximation of a point source. The core diameter of single-mode fiber is around 5 microns.

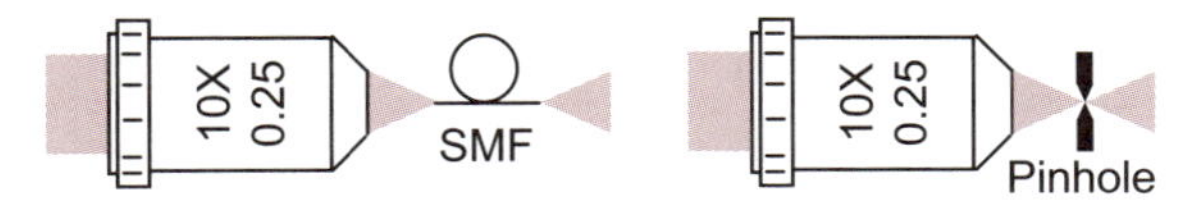

Another approach to spatially filtering a light source is to focus the light onto a small pinhole using a microscope objective. As the magnification of the objective increases, the focused spot size decreases (as does the pinhole size) according to the following equation:

$$D = \frac{1.22\lambda}{NA} \qquad NA = n\sin(\theta)$$

NA is the **numerical aperture** of the objective, where θ is the half angle of the converging beam before the pinhole. The objective creates an 'image' of the source, with the undesirable spatial extent of the source focused off axis. The pinhole blocks all but the central region, and the emerging beam has the spatial coherence properties of a source the size of the pinhole. The chart is for a 1 mm diameter beam at $\lambda = 632.8$ nm.

Objective Magnification	NA	Pinhole Diameter
5X	0.10	50 μm
10X	0.25	25 μm
20X	0.40	15 μm
40X	0.65	10 μm
60X	0.85	5 μm

Polarization

Incoherent light, which does not interfere, adds in irradiance, while coherent light, which does interfere, adds in amplitude. In order for two beams to interfere, the electric fields cannot be completely orthogonal. For example, if the first beam is linearly polarized in x and the second is linearly polarized in y, they will not interfere, regardless of the coherence between the two beams. Two beams with orthogonal polarizations do not interfere, while two beams in the same polarization state have maximum fringe visibility. For polarization states in between, fringe visibility depends on the dot product of the electric field vectors of the two beams. Fringe visibility between two linearly polarized beams goes as $|\cos(\alpha)|$, where α is the angle between the two states of polarization.

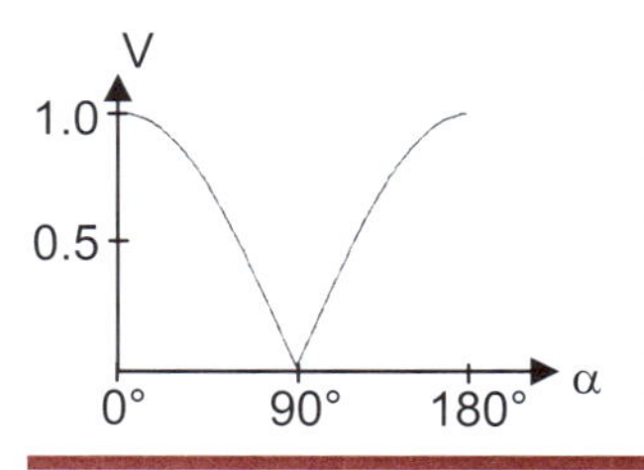

Relative Beam Intensities

Maximum visibility fringes occur when the irradiance of the two beams are equal, which is evident from the two-beam interference equation. As the ratio of the beam intensities deviates from unity, fringe visibility decreases until they are no longer easily detected ($V < 0.2$).

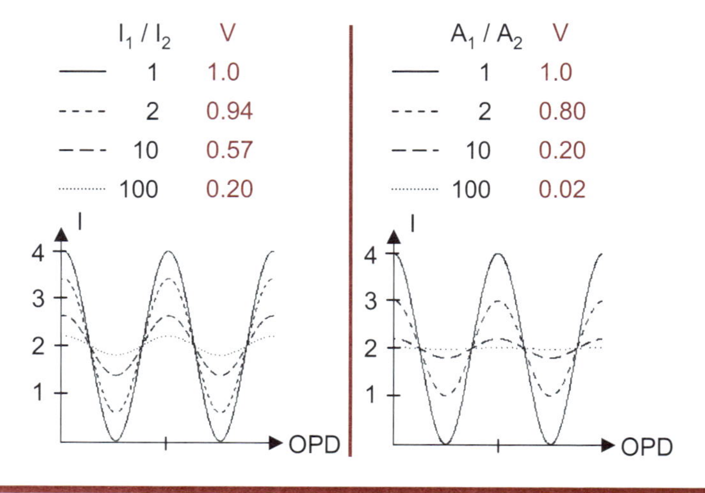

Beamsplitters

Conditions for seeing fringes have been discussed, while methods for creating and recombining two beams have not. Light from one source is split into two beams by a **beamsplitter**, which either divides the wavefront or the amplitude. **Division of wavefront** beamsplitters create two beams from different portions of the original wavefront. A certain degree of spatial coherence is required to see fringes once the beams are recombined since the beams originates from different parts of the wavefront. A classic example of wavefront division is Young's double-slit.

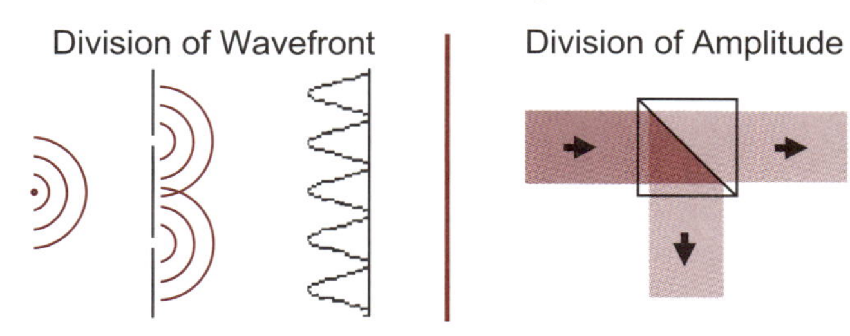

Most laboratory interferometers create two beams using **division of amplitude** beamsplitters, where the irradiance is divided across the entire wavefront such that the diameter of the beam is unchanged. This can be done using cube beamsplitters, plate beamsplitters, pellicles, and diffraction gratings. Light is split as a fraction of incident irradiance in a normal beamsplitter. A **polarization beam splitter**, or **PBS**, splits the light according to its state of polarization, transmitting p-polarized and reflecting s-polarized light. A PBS is one type of **cube beamsplitter**, which in general are made from two right-angle prisms with a coating at the junction of the prisms that gives the desired amount of reflected light. The coating makes it either a PBS or a normal beamsplitter. The external faces of the cube are typically anti-reflection (AR) coated to prevent light loss and spurious fringes. Since light is usually normally incident on the cube face, polychromatic light is not dispersed in a cube beamsplitter.

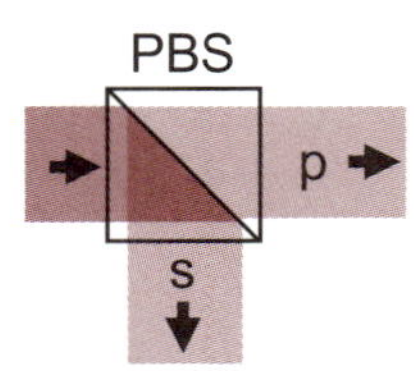

Plate and Pellicle Beamsplitters

Plate beamsplitters are similar to cube beamsplitters in that they divide the amplitude of the incident light and can be made to split the light by polarization or by any desired ratio. One surface is usually AR coated, while the other has the coating to split the beam. Plate beamsplitters can be used to split beams at angles other than 90°. If the source is polychromatic, the index variation in the glass as a function of wavelength will cause a slight spatial displacement between beams of different wavelengths. This is not a concern for sources with a narrow spectrum, such as a laser.

$$d = t\sin(\theta)\left[1 - \sqrt{\frac{1-\sin^2(\theta)}{n^2-\sin^2(\theta)}}\right] \approx t\theta\left(\frac{n-1}{n}\right); \quad \theta \text{ in radians}$$

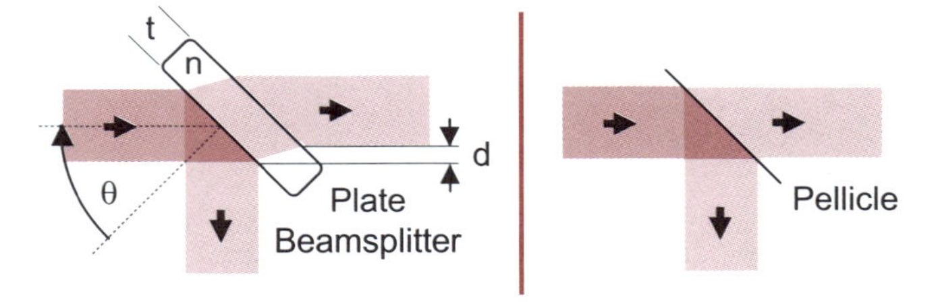

Pellicle beamsplitters are thin (~5 μm) polymers used like a plate beamsplitter and are most useful for beam sampling. They are not susceptible to wavelength displacement in the beam because they are so thin, but they are fragile and very sensitive to vibrations, making them difficult to use in an interferometer.

Diffraction Grating as a Beamsplitter

A diffraction grating can be used as a beamsplitter for a source that is nearly monochromatic. When placed in a beam, part of the beam continues along the original path, while the diffracted beam leaves the grating at an angle θ_d. The grating period is Λ and m is the diffraction order.

$$\sin(\theta_d) - \sin(\theta_i) = \frac{m\lambda}{\Lambda}$$

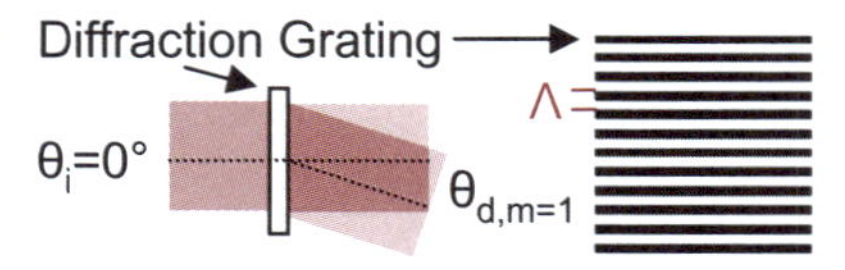

The Interferometer

An interferometer is an instrument that uses the interference of light to make precise measurements of surfaces, thicknesses, surface roughness, optical power, material homogeneity, distances; the list goes on. The resolution of an interferometer is governed by the wavelength of light used and is on the order of a few nanometers. In order to determine the properties of the sample under test, an interferogram is captured and analyzed according to the type of interferometer that created it. Two-beam interferometers return relative information about the OPD between the two beams. Absolute measurements can be made, but extra care must be taken in calibration and system characterization.

Classic Fizeau Interferometer

The classic Fizeau interferometer often uses a spectral emission line to achieve a coherence length of a fraction of a millimeter. The part to be tested is placed on top of a reference flat which has been previously characterized. The **Fresnel reflection** from the bottom of the test piece is the test beam. The reflection from the top surface of the reference flat is the reference beam. The fraction of reflected light ρ at normal incidence is given as a function of the refractive index, n.

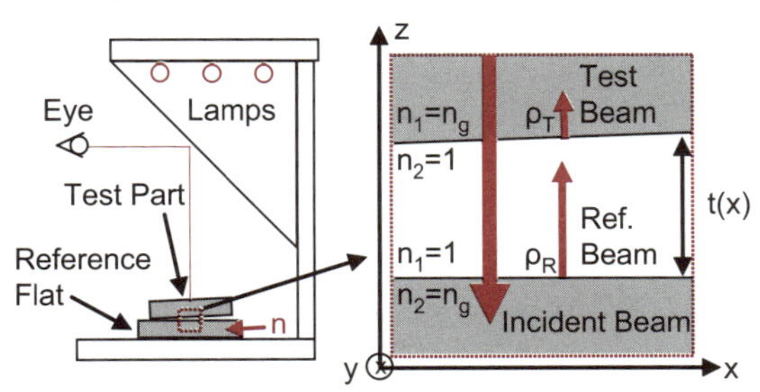

$$\rho = \left(\frac{n_2 - n_1}{n_2 + n_1}\right)^2 \qquad OPD = 2n_{air}t(x) \approx 2t(x); \quad n_{air} \approx 1.0$$

The OPD between the test and reference beams is given above. The reference beam travels the distance between the two surfaces twice; hence the factor of two.

Classic Fizeau Interferograms

The resulting interferogram seen by the eye is shown here for a test part that is tilted in x with respect to the reference part. Any fringe in an interferogram represents a constant OPD. Fringes can be thought of as contour lines on a contour map. Fringes cannot cross each other. The OPD between adjacent bright fringes is one center wavelength of the light source used.

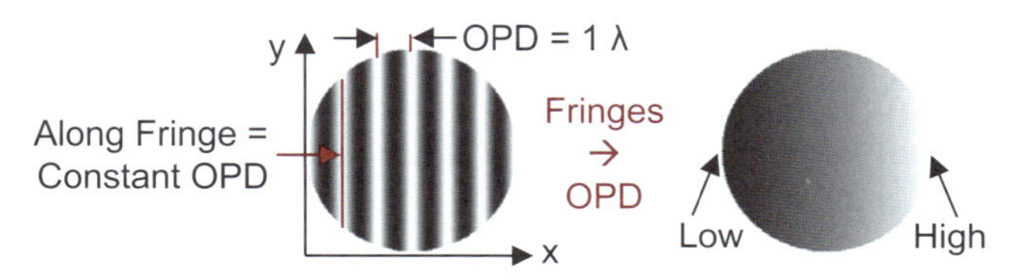

If the test surface has a defect, such as a bump or hole, the fringes will curve around it. To determine if the defect is a bump or a hole on the surface, the direction of increasing OPD must be identified. This can be determined by pushing on one side of the test piece and watching the number of fringes. If the number of fringes increases when pressure is applied, you are pushing on the side with lower OPD. If fringes on the right are higher OPD, meaning the wedge between the two surfaces opens to the right, then this interferogram represents a hole on the bottom surface of the test piece. In the region of the defect, a fringe representing higher OPD is 'pulled in' from the right, indicating that the reference beam traveled further in the region of the defect. Therefore, the defect is a hole with height h, fringe spacing S, and fringe displacement Δ:

$$h = \frac{\lambda}{2}\frac{\Delta}{S}$$

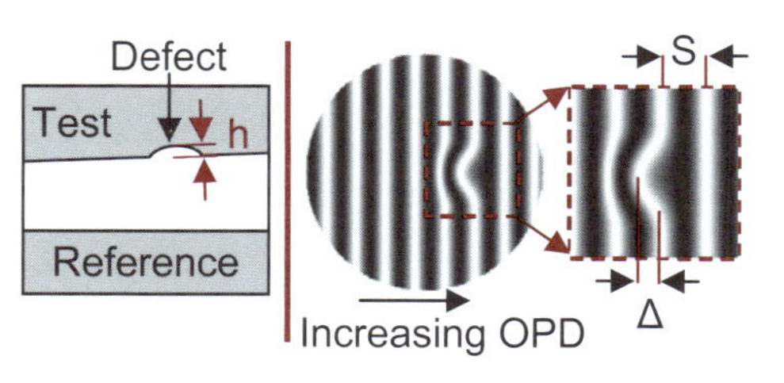

This interferogram is shown with several waves of tilt. Tilt is often purposely introduced so the magnitude of a surface error can be determined. No tilt makes S hard to determine; too much tilt makes Δ hard to find.

Newton's Rings

The classical Fizeau is also useful for testing concave or convex surfaces with large radii of curvature. If the surface has too much curvature the fringe frequency becomes too large and/or the OPD becomes larger than the coherence length. A convex test surface on top of a flat reference will give a **Newton's rings** pattern. If the two surfaces are truly in contact at the center, then the center is always dark in reflection. This is because there is a 180° phase change for the reference beam due to the reflection at a boundary from lower index to higher index. Since the test beam has no phase change, and the OPD is zero, the two waves destructively interfere.

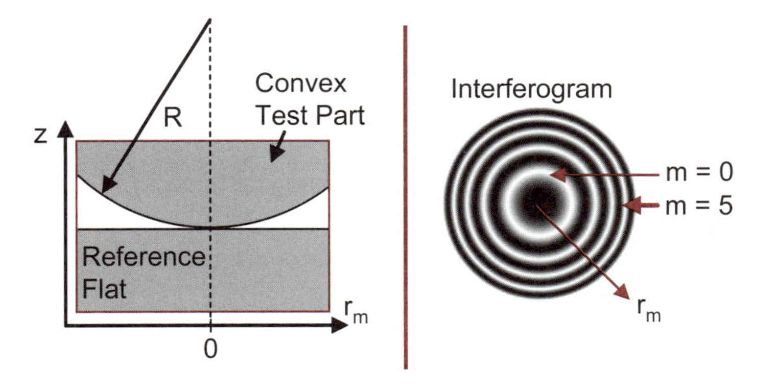

$$R = \frac{r_m^2}{\lambda\left(m + \frac{1}{2}\right)}$$

The radius of curvature R is found by counting the number of bright fringes to a radial distance r_m, where the first bright fringe is $m = 0$.

The classic Fizeau interferometer is useful for checking the curvature of a lens surface versus a master surface, as in lens grinding. The surfaces match when only tilt fringes remain.

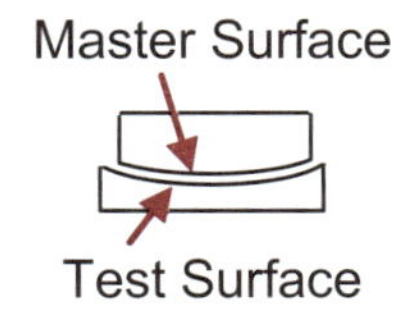

Twyman-Green Interferometer

The **Twyman-Green interferometer** was invented and patented in 1916 and was originally intended for testing prisms and microscope objectives. The invention of the laser increased the utility of the Twyman-Green. Its applicability has grown ever since, and although the **laser-based Fizeau** is probably the most commonly used testing interferometer, the Twyman-Green is useful for introducing important concepts in interferometers.

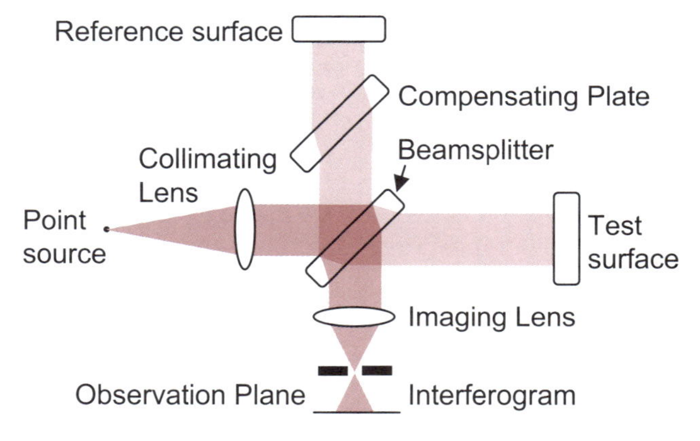

The light source for a Twyman-Green is a quasi-monochromatic point source that is collimated by a collimating lens into a plane wave. This plane wave is split into a reference beam and a test beam by a beamsplitter. The basic setup is for the testing of flats, in which case the reference beam reflects off of a known reference flat and returns to the beamsplitter. The test beam is incident on the unknown test part and also returns to the beamsplitter. The beams are both split a second time, creating two complementary interferograms. One is projected towards the point source, while the more useful interferogram is relayed by an imaging lens to the observation plane.

The image plane is conjugate to the test surface (and reference surface if $OPD = 0$) in order to minimize fringe degradation from diffraction effects.

Compensating Plate

The **compensating plate** is important for light sources with a short coherence length. If the top surface of the beamsplitter is the reflective surface, the test beam passes through the beamsplitter three times before reaching the imaging lens, while the reference beam only passes through it once. In order to match the path length for all wavelengths in glass for the two beams, a compensating plate with a thickness equal to the beamsplitter is placed in the reference arm. The plate is made of the same glass type to match the dispersion for both beams. No compensating plate is necessary if a cube beamsplitter is used.

> If a laser is used as the source, the reference mirror can be placed a different distance from the beamsplitter, creating a Laser Unequal Path Interferometer, or LUPI.

Reflection/Transmission Ratios

Typical non-polarizing beamsplitters will have a reflection to transmission ratio (R/T) of 50/50. Since each beam is reflected and transmitted once before reaching the image plane, at most only half of the total light reaches the image plane. Neglecting losses, 50% is the highest possible fraction of light that can reach the image plane. Any other R/T ratio in a Twyman-Green will return less than 50%.

$$\frac{R}{T}=\frac{0.7}{0.3} \qquad \frac{I_{in}}{I_{out}}=0.7\cdot 0.3+0.3\cdot 0.7=42\%$$

Since the beamsplitter is tilted, it must be larger than the beam so it does not block part of the beam.

PBS-based Twyman-Green Interferometer

If a PBS is used to create the test and reference beams, each arm must include a circular **quarter waveplate** (QWP) oriented at 45° to the s or p linearly polarized light. Since each beam passes through the QWP twice, s-polarized light is converted into p and vice versa. When the light returns to the PBS, the s beam which was initially reflected is now p-polarized and is transmitted to the imaging lens. Similarly, the p-beam is now s-polarized and is reflected.

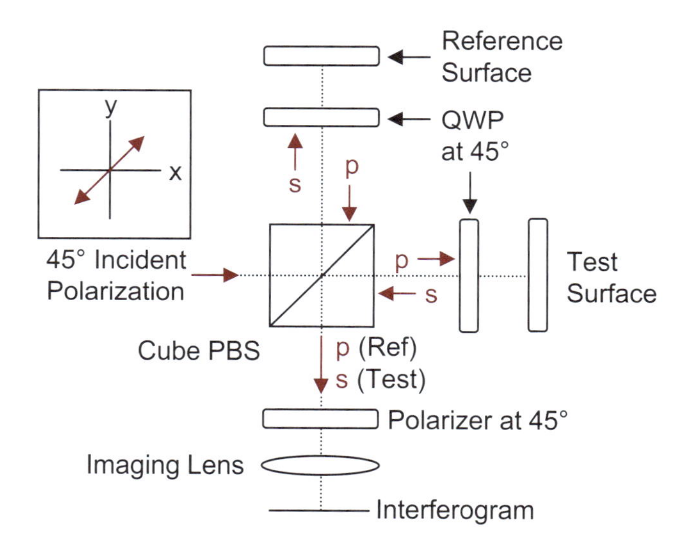

In the plane of observation, the two beams will not interfere because they are in orthogonal polarization states. Therefore, a **polarizer** is included with the transmission axis oriented 45° from either polarization state. This projects the two orthogonal states into the same state. Only half of the light in each polarization state is transmitted, making a PBS-based Twyman-Green just as lossy as the normal setup.

Laser-based Fizeau

Until the invention of the laser, short coherence length sources limited the range of interferometer configurations to those which closely match the path lengths of the two arms of the interferometer. The **laser-based Fizeau** can have large path differences between the two arms due to the long coherence length of lasers. Instead of using a beamsplitter as in the Twyman-Green, the reference beam is typically created from the Fresnel reflection off of a lens or plate surface as shown here. The OPD is then twice the separation of the test surface and the reference surface, which can be as large as a few meters. Single-mode lasers easily satisfy this coherence length requirement.

The biggest advantage of the laser-based Fizeau is that all the optics up to the reference surface are common path. Additionally, it is straightforward to change the system to test different types of optics. The components inside the dashed box remain the same for any test part, while the components outside the box can easily be interchanged. Several commercial interferometers are available that take advantage of this.

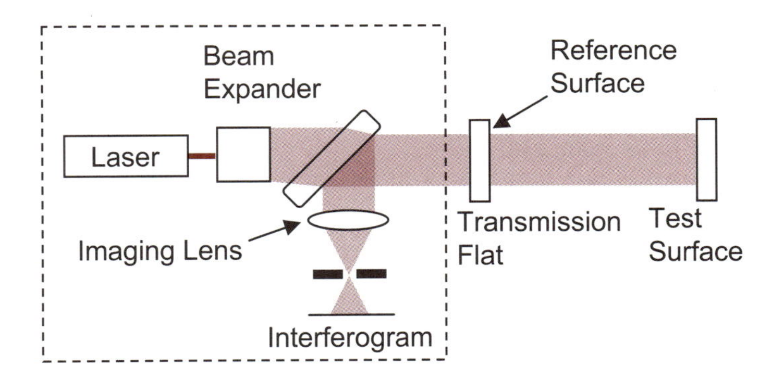

To test a flat, the diverger lens is replaced with a transmission flat while the heart of the interferometer remains unchanged. The reference wavefront comes from the Fresnel reflection off the last surface of the transmission flat.

Laser-based Fizeau (cont'd)

The next setup shown is used for testing concave surfaces. The reference wavefront comes from the Fresnel reflection off the last surface of the diverger lens. The diverger lens is designed so that the beam is normally incident to the reference surface. The flexibility of this interferometer becomes apparent.

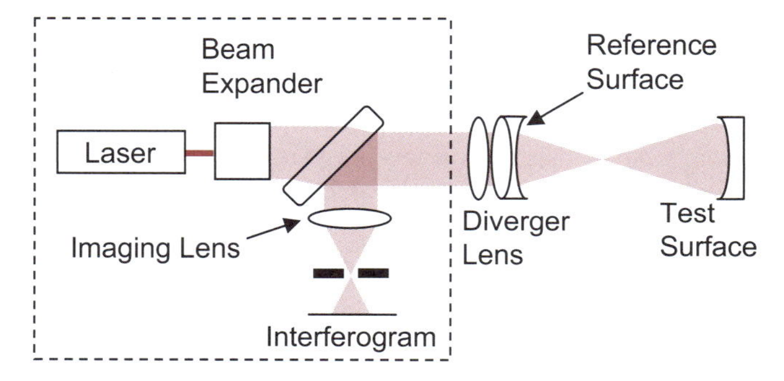

Since the reference beam is created by a Fresnel reflection, the reference beam is about 4% of the incident light. Recall that matching the irradiance in the two interfering beams yields the highest visibility fringes. If the test flat is also uncoated glass, the two beams are well matched. If the test part is a mirror, an attenuator can be inserted in the test beam (after the reference surface) to help match the intensities of the two beams. Many other configurations are used and will be discussed later.

For any setup, the incident beam must be close to normal to the reference surface across the entire beam so that the reference beam retraces the same path back through the interferometer. The same is true for the test beam.

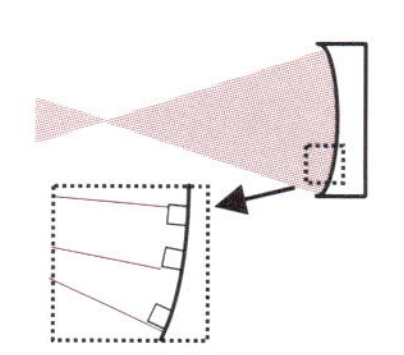

Since the laser source has such a large coherence length, any reflections off of lens surfaces or other components will cause spurious fringes. Much of this is taken care of by the spatial filter at the focus of the imaging lens and diverger lens. These unwanted fringes can be reduced further by adding an anti-reflection coating to the other optical surfaces in the system.

Mach-Zehnder Interferometer

Another common interferometer is the **Mach-Zehnder**, which is useful for testing optics in transmission in single pass. If the optic under test has large deviations, this less sensitive test could be better for keeping the wavefront slope within the dynamic range of the interferometer.

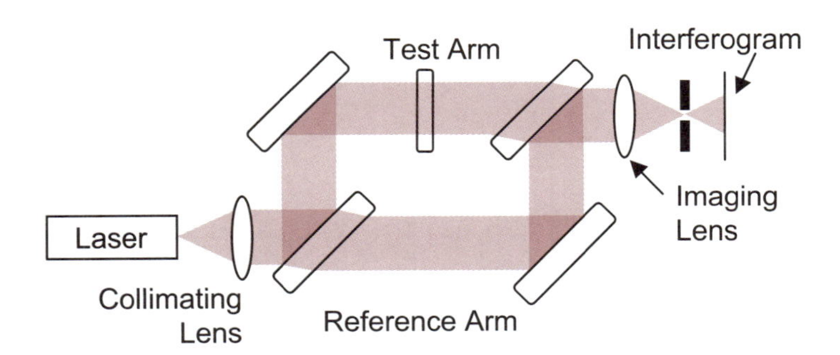

Transmission tests are useful for measuring index homogeneity or thickness variation. In addition, they can be used to measure the power of a lens with low power. The Mach-Zehnder can also be used for testing surfaces in reflection by replacing one of the fold mirrors with the test part. The sensitivity is less since the beam is incident at 45° instead of normal incidence.

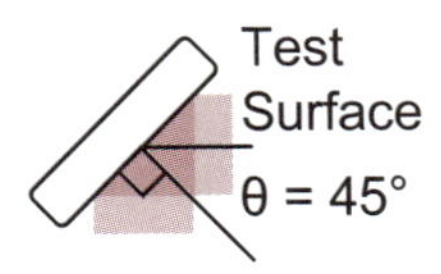

$$\text{Twyman-Green: 1 fringe} = \frac{\lambda}{2} \text{ Surface height error}$$

$$\text{Mach-Zehnder: 1 fringe} = \frac{\lambda}{2 \cdot \cos(45)} = \frac{\lambda}{\sqrt{2}} \text{ Surface error}$$

The imaging lens is chosen such that the test part is conjugate to the plane of the interferogram.

Beam Testing

The Mach-Zehnder interferometer is useful for testing for aberrations in a laser beam. A roughly collimated beam is created by some "black box" system and this input beam needs to be tested for aberration content. The black box system could simply be the collimating optics for a laser.

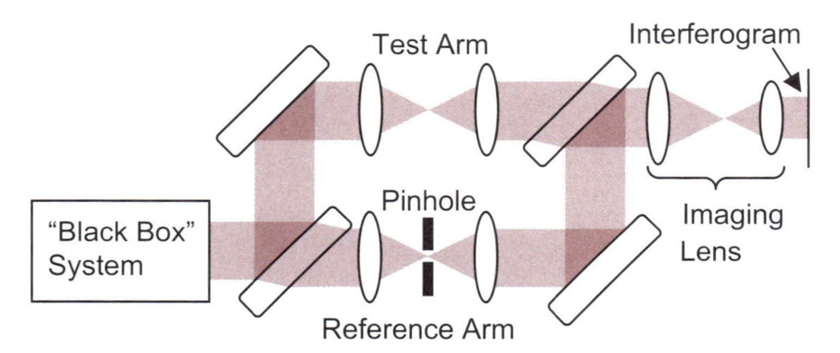

The beam is sent through a Mach-Zehnder interferometer that has two identical pairs of lenses in the two arms which focus and re-collimate the beam. The reference arm has a pinhole at the common focus of the two lenses to block out any light in the input beam that is not collimated. The beam exiting the reference arm is then an aberration-free plane wave. The aberration content of the test beam is unchanged. The two beams are recombined to create an interferogram that reveals the aberration content of the test beam, which is also that of the input beam. The imaging lens determines the part of the beam that is conjugate to the detector, which is the location of the wavefront under test.

Lateral Shear Interferometry

A **lateral shear interferometer** (LSI) interferes the test wavefront $W(x,y)$ with a shifted version of itself. If the shift (shear) is Δx in the x direction, a bright fringe appears for the following condition (m is an integer):

$$W(x+\Delta x, y) - W(x,y) = m\lambda = \left[\frac{\delta W(x,y)}{\delta x}_{\text{Avg. over Shear}}\right]\Delta x$$

So a lateral shear interferogram shows the average local wavefront slope along the direction of the shear. There are many different lateral shear interferometers. The **Murty plane-plate lateral shear interferometer** is useful for working with collimated light that has a long temporal coherence length compared with the plate thickness. The amount of shear is controlled by the tilt and thickness of the plate. Often, wedge is built into the plate in the direction orthogonal to the shear (y) to give straight line fringes when the light is collimated.

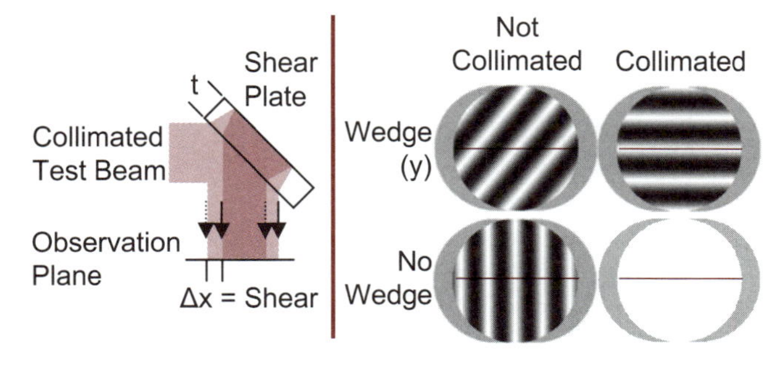

Another common method for lateral shear is to use a two-frequency diffraction grating. The grating, placed near the focus of the lens under test, creates two diffracted beams. The shear comes from the slightly different diffraction angles, and the resulting interferogram shows the local slope of the wavefront, not the wavefront shape.

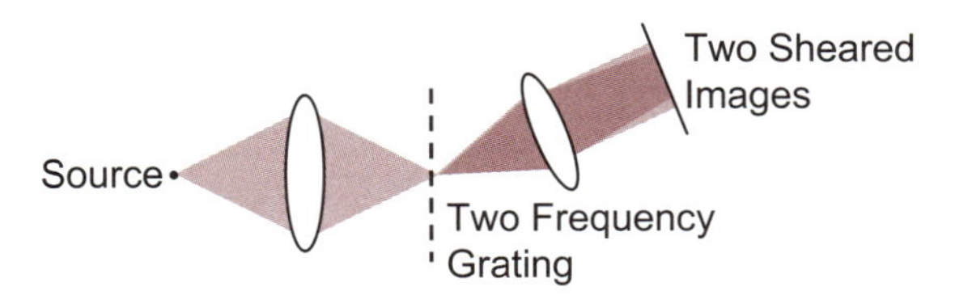

Rotating Grating LSI

Two crossed gratings with the same spatial frequencies can be put together to form a lateral shear interferometer. As one grating is rotated an angle α, the first diffraction orders from the two gratings move across each other, increasing the shear. This ability to easily vary the shear is useful for testing wavefronts with much aberration.

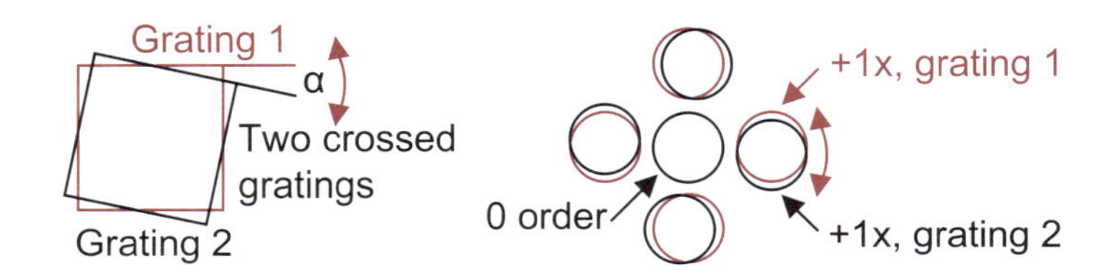

Lateral shear can also be introduced with a **Savart plate**. Two identical uniaxial crystal plates with their optic axis cut at 45° to the plate normal are combined. This creates two parallel sheared copies of the input wavefront. If the two indices of refraction are n_o and n_e, the shear is:

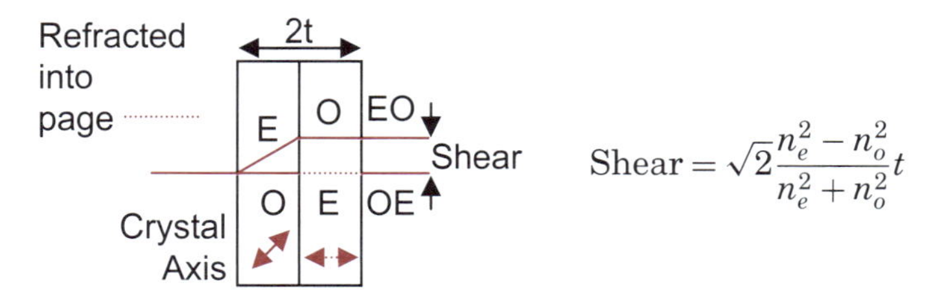

$$\text{Shear} = \sqrt{2}\frac{n_e^2 - n_o^2}{n_e^2 + n_o^2}t$$

A **Wollaston prism** combines two pieces of uniaxial crystal such that an angular shear is achieved via a polarization-dependent refraction at the interface of the two crystals. If the wedge angle is α, the shear angle, θ, is given here:

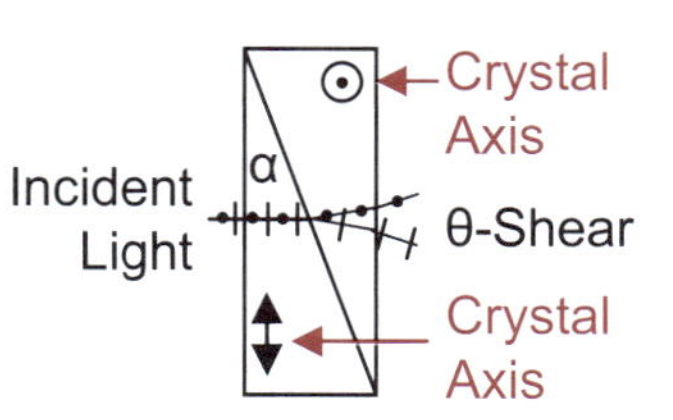

$$\theta = 2(n_e - n_o)\tan\alpha$$

Radial Shear Interferometer

A radial shear interferometer introduces radially symmetric shear by interfering the wavefront with an expanded copy of itself. A large shear gives similar results to a Twyman-Green interferometer, while small shear is a low-sensitivity, high-dynamic-range test.

$$\text{Radial Shear} = R_s = \frac{S_1}{S_2}$$

The resulting interferogram is the same as a Twyman-Green if each aberration coefficient is divided by $(1 - R_s^n)$, where n is the radial power of each particular aberration.

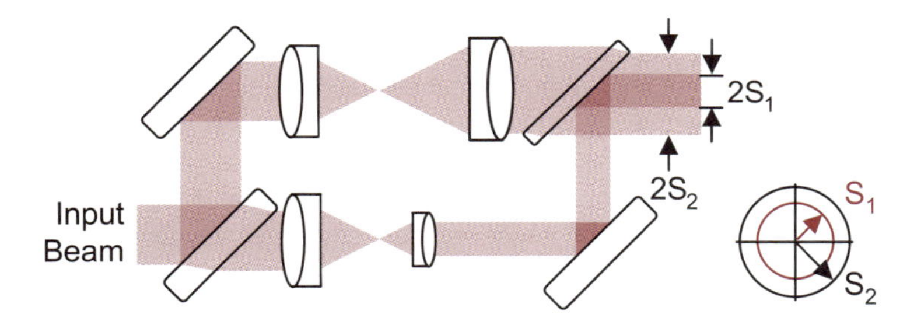

Interferograms

In addition to revealing local defects in a surface or part, interferograms contain information about the **aberration content** of the test wavefront, which relates directly to aberrations in the test part. It is up to the interpreter to decide the source of any aberrations in the interferogram based on the type and configuration of interferometer used. It is important to remember that interferograms represent the *difference* between the two interfering beams. Therefore, aberrations due to anything that both beams go through, such as the imaging lens or the beam expander, do not change the interferogram.

In this context, aberrations refer to the wavefront errors introduced that are inherent to rotationally symmetric optical systems. These aberrations are variations of the actual wavefront from an ideal spherical wavefront. In interferometers, the ideal spherical wavefront is often a plane wave; a spherical wavefront with an infinite radius of curvature. The wavefront errors derived from the interferogram are the errors in the measurement plane, which is the plane conjugate to the detector plane via the imaging lens.

$$\frac{1}{z'} = \frac{1}{z} + \frac{1}{f}$$

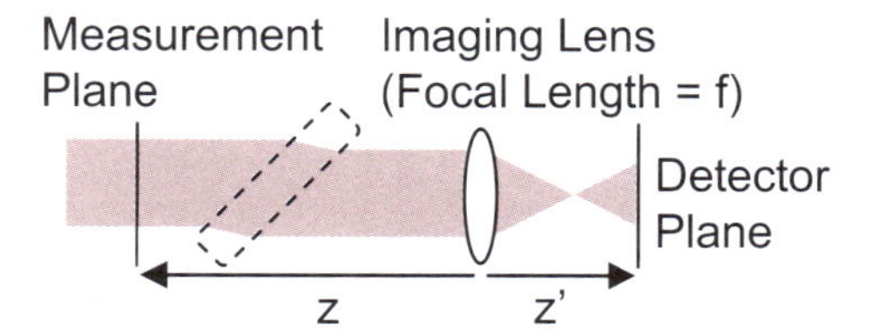

There are two common representations of these aberrations which will be described here: **wavefront aberration coefficients** and **Zernike polynomials**. Both are limited in that they are only capable of accurately representing the aberrations inherent to rotationally symmetric optical systems. Local defects, air turbulence, index variations, and fabrication errors are not likely to be well described by these representations.

Wavefront Aberration Coefficients

These terms come from a power series expansion of the wavefront aberrations in a rotationally symmetric optical system. The pupil is normalized to the unit circle with points in that circle represented by ρ and θ. Note that θ is measured *clockwise* from the y_P axis.

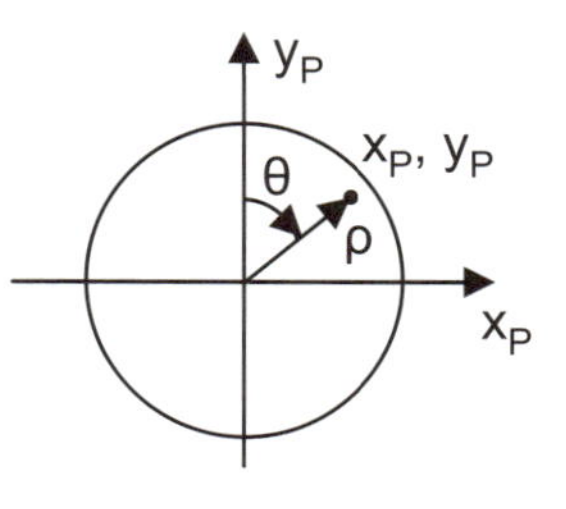

$$x_P = \rho \sin\theta \quad y_P = \rho\cos\theta \quad \rho = \sqrt{x_P + y_P}$$

The wavefront coefficients are W_{IJK}, where the subscripts tell the powers of the three expansion terms: H, the normalized field coordinate; ρ, normalized distance from the optical axis in the measurement plane; and $\cos\theta$.

$$W = \sum_{L,K,N} W_{IJK} H^I \rho^J \cos^K\theta; \quad I = 2L + K, J = 2N + K$$

	First-Order Terms
$W = W_{020}\rho^2$	**Defocus**
$+ W_{111} H\rho\cos\theta$	**Wavefront Tilt**
	Third-Order Terms
$+ W_{040}\rho^4$	**Spherical Aberration**
$+ W_{131} H\rho^3\cos\theta$	**Coma**
$+ W_{222} H^2\rho^2\cos^2\theta$	**Astigmatism**
$+ W_{220} H^2\rho^2$	**Field Curvature**
$+ W_{311} H^3\rho\cos\theta$	**Distortion**
	Higher-Order Terms
$+ \cdots$	

Since H and ρ are normalized, the W coefficients will change if the size of the test plane is changed. The W coefficients can be given in units of length or in a unitless value of the number of wavelengths of each aberration. An interferogram measures a single field point, H. Therefore, field curvature can look like defocus and distortion can look like tilt. To fully determine the wavefront aberration coefficients, a number of field points must be measured.

Zernike Polynomials

Zernike polynomials were first derived by Fritz Zernike in 1934. They are useful in expressing wavefront data since they are of the same form as the types of aberrations often observed in optical tests. These polynomials are a complete set in two variables, ρ and θ', that are orthogonal in a continuous fashion over the unit circle. It is important to note that Zernikes are orthogonal only in a continuous fashion and that in general they will not be orthogonal over a discrete set of data points. Note that θ' is the angle *counterclockwise* from the x_P axis.

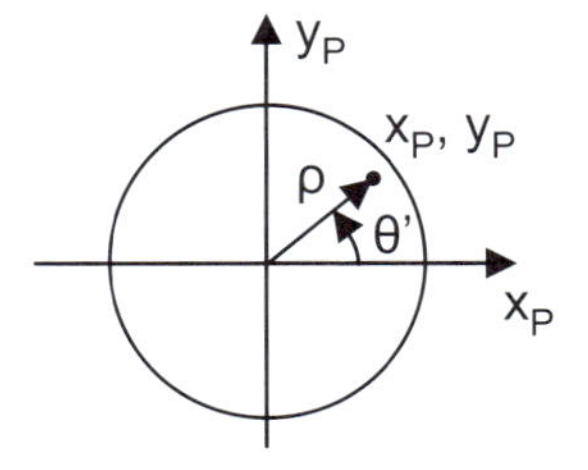

$$x_P = \rho\cos\theta' \quad y_P = \rho\sin\theta' \quad \rho = \sqrt{x_P + y_P}$$

There are several common definitions for the Zernike polynomials, so care should be taken that the same set is used when comparing Zernike coefficients. Also note the difference in the definitions of the angle in the measurement plane, θ vs. θ'.

One of the convenient features of Zernike polynomials is that their simple rotational symmetry allows the polynomials to be expressed as products of radial terms and functions of angle, $r(\rho)g(\theta')$, where $g(\theta')$ is continuous and repeats itself every 2π radians. The coordinate system can be rotated by an angle α without changing the form of the polynomial. Each Zernike term is referenced by a single number or by two subscripts, n and m, where both are positive integers or zero.

$$g(\theta' + \alpha) = g(\theta')g(\alpha) \quad g(\theta') = e^{\pm im\theta'}$$

The radial function, $r(\rho)$, must be a polynomial in ρ of degree n and contain no power of ρ less than m. Also, $r(\rho)$ must be even if m is even and odd if m is odd.

Zernike Polynomials (cont'd)

n	m	$Z_{\#}$	Polynomial	Aberration
0	0	0	1	***Piston***
1	1	1	$\rho\cos\theta'$	***Tilt x***
		2	$\rho\sin\theta'$	***Tilt y***
	0	3	$2\rho^2-1$	***Defocus, Piston***
2	2	4	$\rho^2\cos 2\theta'$	***Astigmatism, Defocus***
		5	$\rho^2\sin 2\theta'$	***Astigmatism, Defocus***
	1	6	$(3\rho^2-2)\rho\cos\theta'$	***Coma, Tilt x***
		7	$(3\rho^2-2)\rho\sin\theta'$	***Coma, Tilt y***
	0	8	$6\rho^4-6\rho^2+1$	***Spherical, Defocus***
3	3	9	$\rho^3\cos 3\theta'$	***Higher-Order Terms*** ...
		10	$\rho^3\sin 3\theta'$	
	2	11	$(4\rho^2-3)\rho^2\cos 2\theta'$	
		12	$(4\rho^2-3)\rho^2\sin 2\theta'$	
	1	13	$(10\rho^4-12\rho^2+3)\rho\cos\theta'$	
		14	$(10\rho^4-12\rho^2+3)\rho\sin\theta'$	
	0	15	$20\rho^6-30\rho^4+12\rho^2-1$	
4	4	16	$\rho^4\cos 4\theta'$	
		17	$\rho^4\sin 4\theta'$	
	3	18	$(5\rho^2-4)\rho^3\cos 3\theta'$	
		19	$(5\rho^2-4)\rho^3\sin 3\theta'$	
	2	20	$(15\rho^4-20\rho^2+6)\rho^2\cos 2\theta'$	
		21	$(15\rho^4-20\rho^2+6)\rho^2\sin 2\theta'$	
	1	22	$(35\rho^6-60\rho^4+30\rho^2-4)\rho\cos\theta'$	
		23	$(35\rho^6-60\rho^4+30\rho^2-4)\rho\sin\theta'$	
	0	24	$70\rho^8-140\rho^6+90\rho^4-20\rho^2+1$	
5	5	25	$\rho^5\cos 5\theta'$	
		26	$\rho^5\sin 5\theta'$	
	4	27	$(6\rho^2-5)\rho^4\cos 4\theta'$	
		28	$(6\rho^2-5)\rho^4\sin 4\theta'$	
	3	29	$(21\rho^4-30\rho^2+10)\rho^3\cos 3\theta'$	
		30	$(21\rho^4-30\rho^2+10)\rho^3\sin 3\theta'$	
	2	31	$(56\rho^6-105\rho^4+60\rho^2-10)\rho^2\cos 2\theta'$	
		32	$(56\rho^6-105\rho^4+60\rho^2-10)\rho^2\sin 2\theta'$	
	1	33	$(126\rho^8-280\rho^6+210\rho^4-60\rho^2+5)\rho\cos\theta'$	
		34	$(126\rho^8-280\rho^6+210\rho^4-60\rho^2+5)\rho\sin\theta'$	
	0	35	$252\rho^{10}-630\rho^6+560\rho^6-210\rho^4+30\rho^2-1$	

RMS Wavefront Error

Each Zernike term contains the appropriate amount of each lower-order term to make it orthogonal to each lower-order term. The average value of each term over the unit circle is zero. Also, each term minimizes the root-mean-square (RMS) wavefront error to the order of that term, where **RMS wavefront error** is σ and wavefront variance is σ^2:

$$\sigma^2 = \frac{1}{\pi}\int_0^{2\pi}\int_0^1 [\Delta W(\rho,\theta) - \overline{\Delta W}]^2 \rho d\rho d\theta = \overline{\Delta W^2} - (\overline{\Delta W})^2$$

$\overline{\Delta W}$ is the mean wavefront OPD and $\Delta W(\rho,\theta)$ is the point-by-point error measured relative to the desired wavefront, which is usually the best fit spherical wave. The RMS is usually a better representation of the overall wavefront error than the peak-to-valley (P-V), which is the sum of the absolute values of the maximum departure in both positive and negative directions.

Interferograms are measurements of the wavefront aberrations at a single field position. It is possible to convert between the Zernike coefficients at a given field and the field-independent aberration terms.

Term	Magnitude	Angle
$W_{(1)11}$	$\sqrt{(Z_1 - 2Z_6)^2 + (Z_2 - 2Z_7)^2}$	$\tan^{-1}\left(\frac{Z_2 - 2Z_7}{Z_1 - 2Z_6}\right)$
$W_{(0)20}$	$2Z_3 - 6Z_8 \pm \sqrt{Z_4^2 + Z_5^2}$	N/A
$W_{(2)22}$	$\mp 2\sqrt{Z_4^2 + Z_5^2}$	$\frac{1}{2}\tan^{-1}(Z_5/Z_4)^*$
$W_{(1)31}$	$3\sqrt{Z_6^2 + Z_7^2}$	$\tan^{-1}(Z_7/Z_6)$
$W_{(0)40}$	$6Z_8$	N/A

The field dependence subscript is given in parentheses to identify each aberration. The sign chosen for $W_{(0)20}$ is chosen to minimize the absolute value of its magnitude. The sign for $W_{(2)22}$ is opposite to the sign chosen for $W_{(0)20}$.

*If the sign chosen for $W_{(0)20}$ is negative, add $\pi/2$ radians to the angle for $W_{(2)22}$.

Spherical Aberration Interferograms

The following figures are example interferograms with their corresponding wavefront aberration and Zernike coefficients in waves.

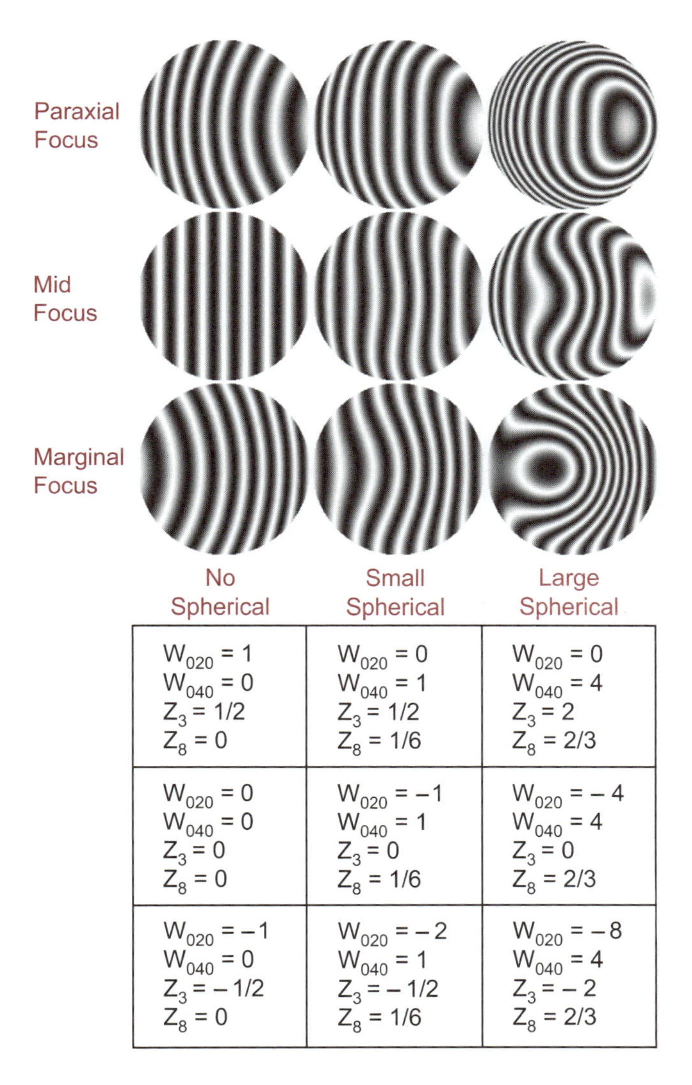

No Spherical	Small Spherical	Large Spherical
$W_{020} = 1$ $W_{040} = 0$ $Z_3 = 1/2$ $Z_8 = 0$	$W_{020} = 0$ $W_{040} = 1$ $Z_3 = 1/2$ $Z_8 = 1/6$	$W_{020} = 0$ $W_{040} = 4$ $Z_3 = 2$ $Z_8 = 2/3$
$W_{020} = 0$ $W_{040} = 0$ $Z_3 = 0$ $Z_8 = 0$	$W_{020} = -1$ $W_{040} = 1$ $Z_3 = 0$ $Z_8 = 1/6$	$W_{020} = -4$ $W_{040} = 4$ $Z_3 = 0$ $Z_8 = 2/3$
$W_{020} = -1$ $W_{040} = 0$ $Z_3 = -1/2$ $Z_8 = 0$	$W_{020} = -2$ $W_{040} = 1$ $Z_3 = -1/2$ $Z_8 = 1/6$	$W_{020} = -8$ $W_{040} = 4$ $Z_3 = -2$ $Z_8 = 2/3$

$W_{111} = -4$ and $Z_1 = -4$ for all of these interferograms.

Astigmatism Interferograms

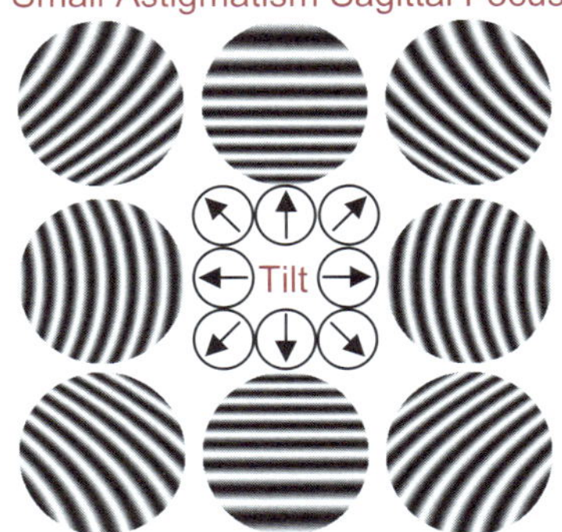

$\theta =$ Tilt orientation

Wavefront Coefficients

$W_{020} = 0$

$W_{111} = 4$ at θ

$W_{222} = -1$ at 90°

Zernike Coefficients

$Z_0 = -\frac{1}{4}$ $\quad Z_3 = -\frac{1}{4}$

$Z_1 = 4\cos\theta$ $\quad Z_4 = \frac{1}{2}$

$Z_2 = 4\sin\theta$ $\quad Z_5 = 0$

Large Astigmatism Sagittal Focus

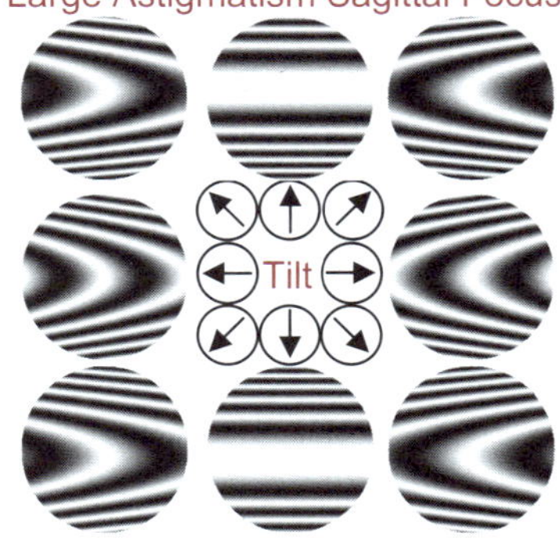

Wavefront Coefficients

$W_{020} = 0$

$W_{111} = 1$ at θ

$W_{222} = -4$ at 90°

Zernike Coefficients

$Z_0 = -1$ $\quad Z_3 = -1$

$Z_1 = \cos\theta$ $\quad Z_4 = 2$

$Z_2 = \sin\theta$ $\quad Z_5 = 0$

Large Astigmatism Medial Focus

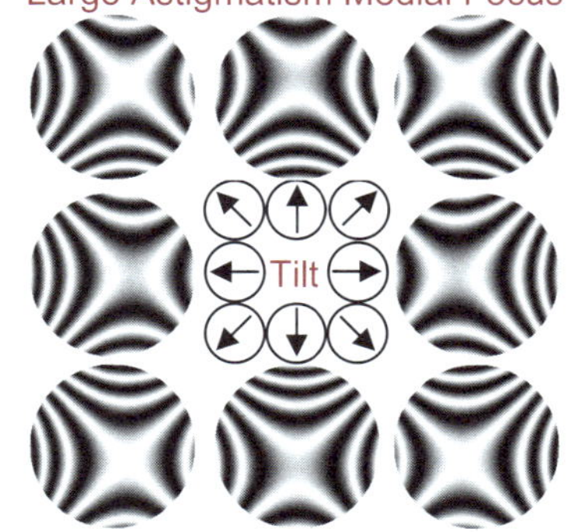

Wavefront Coefficients

$W_{020} = 2$

$W_{111} = 1$ at θ

$W_{222} = -4$ at 90°

Zernike Coefficients

$Z_0 = 0$ $\quad Z_3 = 0$

$Z_1 = \cos\theta$ $\quad Z_4 = 2$

$Z_2 = \sin\theta$ $\quad Z_5 = 0$

Interferograms—Other Aberrations

Large Coma, Varying Tilt

W_{111}: −4 at 90°; −3 at 112.5°, −3 at 90°, −3 at 67.5°; −2 at 90°

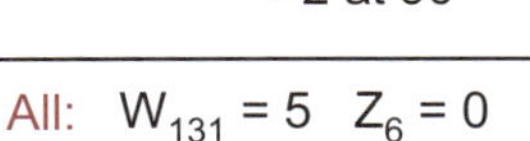

All: $W_{131} = 5$, $Z_6 = 0$, $Z_7 = 5/3$

Z_1, Z_2: 0, −2/3; 1.148, 0.562; 0, 1/3; −1.148, 0.562; 0, 4/3

Aberration Variation by Changing Defocus

$W_{040} = 4$, $Z_8 = 2/3$

$Z_0 = 5/6$, 4/3, 11/6
$Z_3 = 3/2$, 2, 5/2

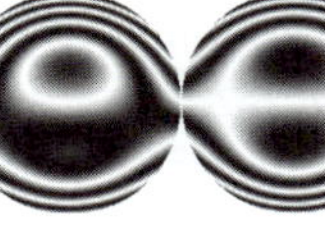
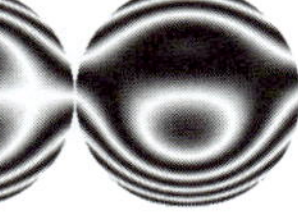

$W_{111} = 2$, $W_{131} = -5$, $Z_2 = -4/3$, $Z_7 = 5/3$

$Z_0 = -1/2$, 0, 1/2
$Z_3 = -1/2$, 0, 1/2

$W_{020} =$ −1, 0, 1, 2, 3

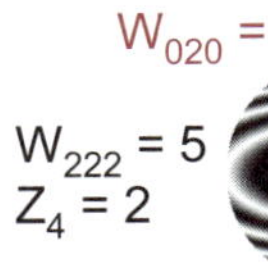
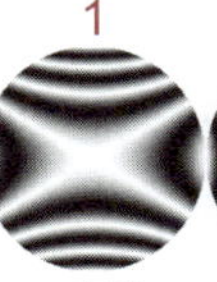

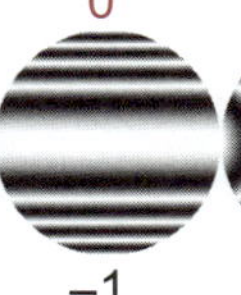

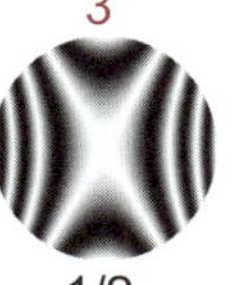

$W_{222} = 5$, $Z_4 = 2$

$Z_0 = Z_3 =$ −3/2, −1, −1/2, 0, 1/2

Sagittal (at −1); Medial (at 0)

Combined Aberrations

Spherical; Sph + Coma

Coma; Sph + Astig; Sph + Coma + Astig

Astigmatism; Coma + Astig

Moiré

In optics, **moiré** refers to a beat pattern produced between two gratings of approximately equal spacing. Examples of moiré can be seen using household items, such as overlapping two window screens or with a striped shirt seen on television. In 1874, Lord Rayleigh pioneered the use of moiré for reduced sensitivity testing. Moiré is useful to help understand basic interferometry.

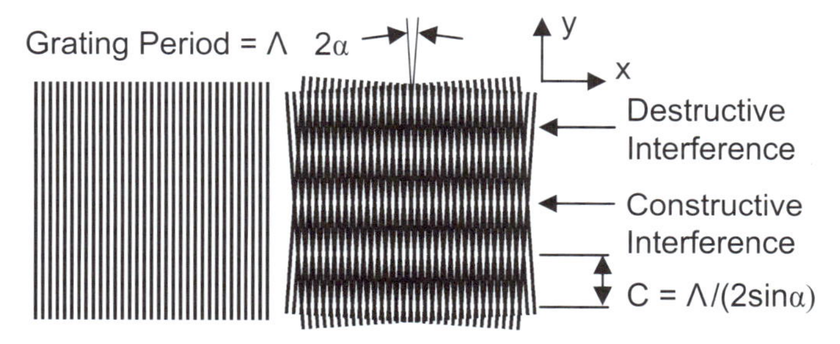

The basic idea in using moiré for optical testing is to project a fringe pattern or grating onto an object and then view it from a different angle. Surface topography can be recovered from the resulting moiré pattern. This technique is capable of contouring surfaces that are too coarse or have too much deviation for traditional interferometry. Mathematically, the moiré pattern is the product of the two gratings. In general, the fringe spacing C in the moiré pattern from two straight-line gratings of period Λ_1 and Λ_2, oriented 2α from each other is given by:

$$C = \frac{\Lambda_1 \Lambda_2}{\sqrt{\Lambda_2^2 \sin^2 2\alpha + (\Lambda_2 \cos 2\alpha - \Lambda_1)^2}}$$

In the above figure, $\Lambda_1 = \Lambda_2 = \Lambda$. On the left, $\alpha = 0$ but $\Lambda_1 \neq \Lambda_2$, so the fringe spacing equals the beat wavelength.

$$C = \Lambda_{beat} = \frac{\Lambda_1 \Lambda_2}{|\Lambda_2 - \Lambda_1|}$$

On the right, the gratings are tilted and $\Lambda_1 \neq \Lambda_2$. If Λ_1 is known and C is measured, then Λ_2 and α can be found.

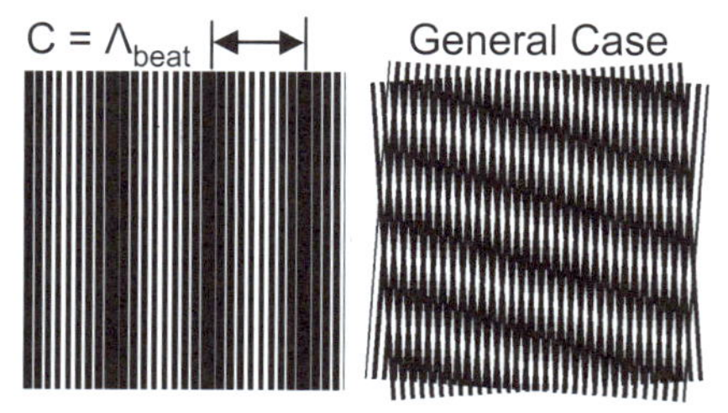

Moiré and Interferograms

The single grating in the figure below can be thought of as a snapshot of a plane wave traveling in the $+x$ direction with the grating period Λ replaced with the wavelength of light, λ. The moiré pattern on the right can be thought of as two plane waves from an interferometer traveling in the $+x$ direction with an angle of 2α, frozen in time. Since $\lambda_1 = \lambda_2$, the regions of destructive and constructive interference remain stationary in y. The moiré from two straight line gratings correctly predicts the centers of the interference fringes produced by interfering two plane waves (tilt fringes) but not the sinusoidal fringe pattern.

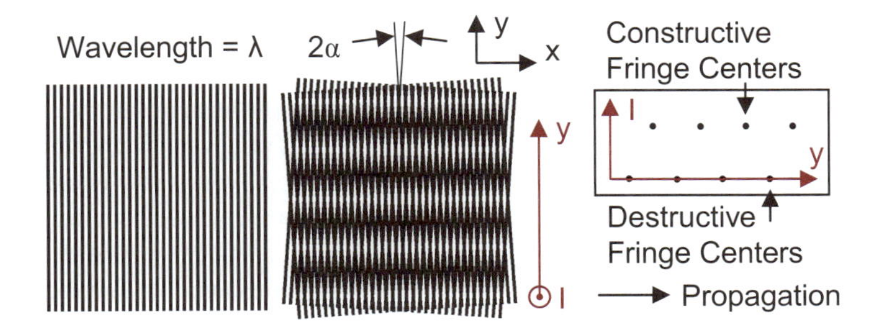

The moiré pattern resulting from superimposing (multiplying) two interferograms (a and b) shows the difference in the aberrations between the two (c).

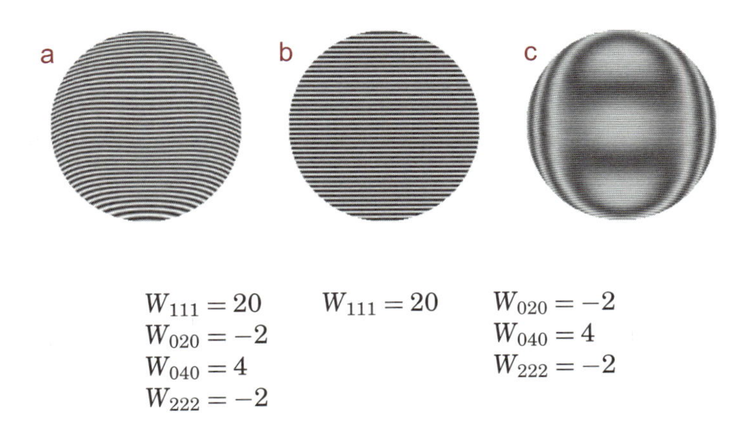

Direct Phase Measurement

Digitization of single interferograms into a computer using a camera was an important step in increasing the resolution of interferometry. The information contained in the fringe centers of a single interferogram describes the shape of the surface, but more information is needed to determine the sign. In addition, determining fringe centers is sensitive to intensity variations, detector nonuniformity and pixel spacing.

Phase-shifting interferometry, or PSI, is a method that records a series of interferograms where the phase of one of the two interfering beams is changed by a known amount and direction between images. The wavefront phase, including its sign, can be found from the variation of intensity at each pixel between images. The improvements over static interferometry include much higher measurement accuracy (better than 1/1000 fringe) and good results from low-contrast fringes. Intensity variations across the pupil, either from pixel to pixel nonuniformity or source distribution, no longer matter since the phase is calculated at each pixel independently. The result is a phase map at a fixed grid of points. For example, a Twyman-Green interferometer testing a surface in reflection with height errors h using a light source with wavelength λ will have a wavefront phase ϕ:

$$\phi(x, y) = 4\pi h(x, y)/\lambda$$

$$I(x, y) = I_{dc}(x, y) + I_{ac}(x, y)\cos[\phi(x, y) + \phi(t)]$$

The irradiance I is measured at each pixel for each image. I_{dc} and I_{ac} and $\phi(x, y)$ are all unknowns, so at least three interferograms are needed to determine $\phi(x, y)$. The phase change between images is $\phi(t)$. The most common phase shift between images is $\pi/2$, or 90°, shown in these four images with tilt fringes.

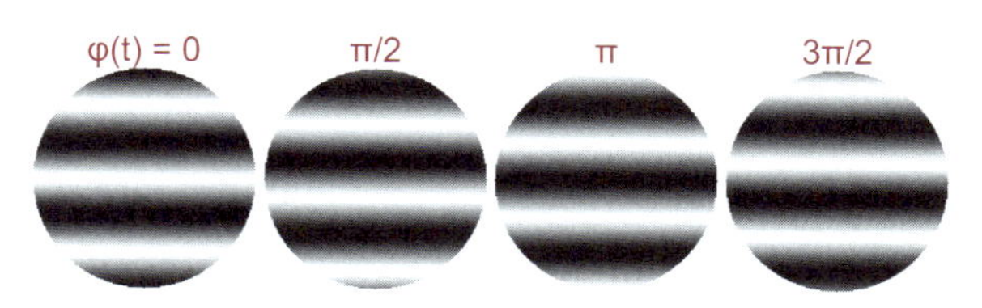

Methods for Phase Shifting

In order to create multiple interferograms, a method is needed for introducing the proper phase shift. The most straightforward and by far the most common method for changing the phase is a movable mirror. A phase shift of 90° in a Twyman-Green interferometer is equivalent to moving the reference mirror $\lambda/8$. For a HeNe laser at 632.8 nm, this requires moving the mirror about 79.1 nm for each image. **Piezoelectric transducers** (PZTs) are quite common in phase shifting interferometers. A DC voltage, typically of a few hundred volts, is used to expand or contract the PZT material and create a small position change. The same method can be used in a Mach-Zehnder except the distance the mirror travels is increased by a factor of $1/\cos(\theta)$. This also introduces a small lateral displacement, which is ignored. In a laser-based Fizeau, either the test piece or the entire reference objective is translated.

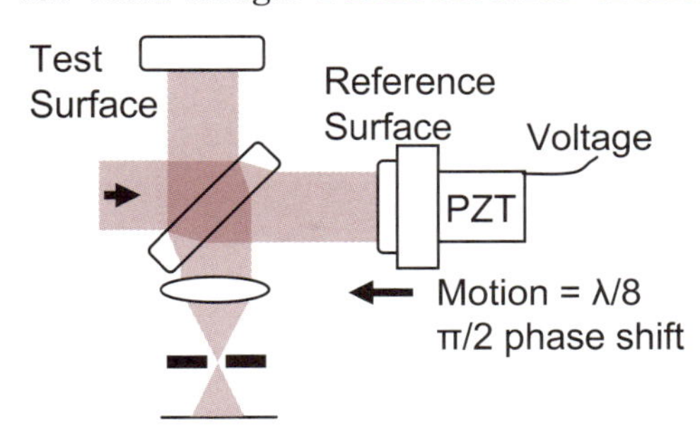

Another method for phase shifting by changing the path length is to rotate a plane parallel plate in the collimated reference beam. **Snell's Law**

$$n_1 \sin\theta = n_2 \sin\theta'$$

shows that the beam will travel through more glass as θ increases, causing an increase in OPL and a nonlinear phase shift. A similar effect is achieved by translating a prism orthogonal to the beam. Since both methods are functions of the index of refraction, n, they are not achromatic. In addition, both methods are only useful in collimated beams due to aberrations induced for a converging or diverging beam.

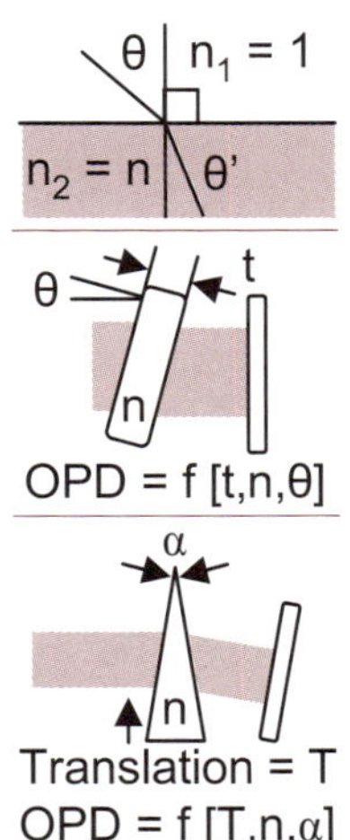

Continuous Phase Shifting

Continuous phase shifting can be achieved by changing the frequency of the light in one arm of the interferometer by $\Delta\nu$. The phase at each point in the interferogram will change with time, or $\phi(t) = 2\pi\Delta\nu t$. The frequency shift can come from a translating diffraction grating, an acousto-optic (AO) **Bragg cell**, or rotating polarization phase retarders.

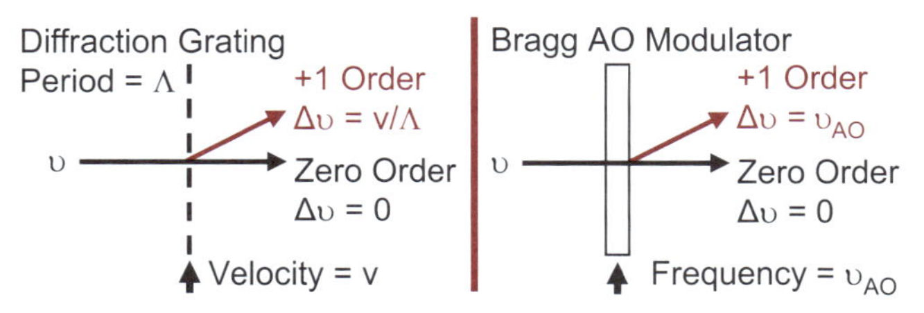

Sending a circularly polarized beam through a half-wave plate rotating at a rate Ω changes the frequency of the light by 2Ω.

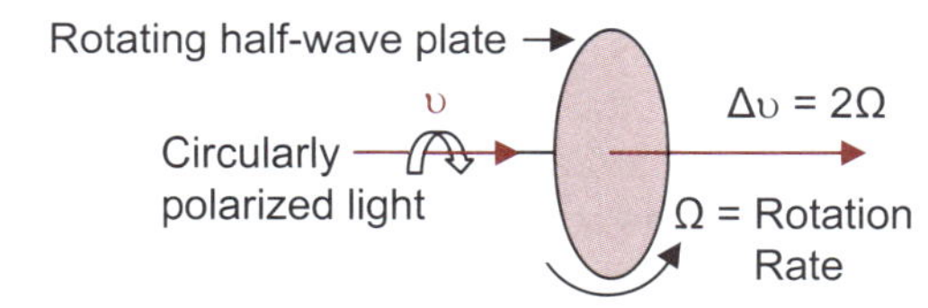

A polarizer can also be used to create phase shifts by changing the OPD. If the reference beam is left-hand circular polarized (LHC) and the test beam is right-hand circular (RHC), they can be interfered using a linear polarizer. The total phase between the two beams is that from the test part, $\phi(x, y)$ and the phase shift, which is twice the angle of the transmission axis of the polarizer (θ). Changing the orientation of the polarizer changes the phase shift between the two beams.

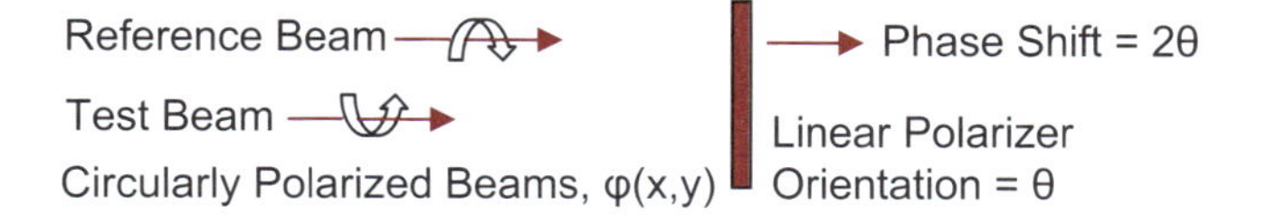

Rotating the polarizer continuously changes the frequency between the two beams.

Liquid Crystal Retarder

Another continuous phase shifting method using orthogonal polarizations has the input beam linearly polarized at 45° incident on a **liquid crystal retarder** at 0°. As the retardance is changed, the phase of the linearly polarized beam at 0° is changed relative to the beam at 90°. The beams are split by a PBS. Another polarizer is used near the detector to recombine the beams so they interfere.

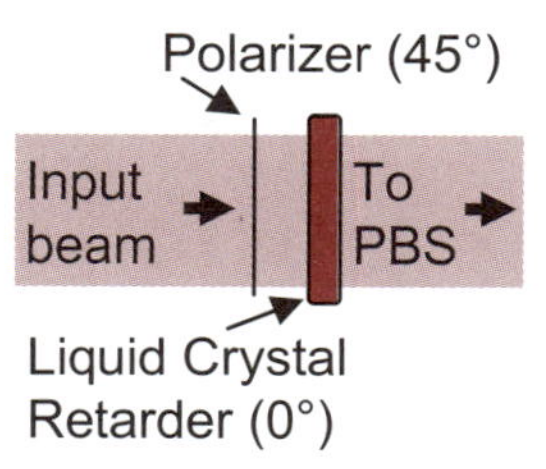

Frequency Shifting Source

This method utilizes a diode laser, tunable laser or other source that can tune the frequency of the light emitted. The phase shift $\Delta\phi$ introduced by changing the optical frequency by $\Delta\nu$ depends on the OPD between the test and reference beams. No moving parts are required.

$$\phi = \nu \frac{2\pi}{c}(2 \cdot OPD)$$

$$\Delta\phi = \Delta\nu \frac{2\pi}{c}(2 \cdot OPD)$$

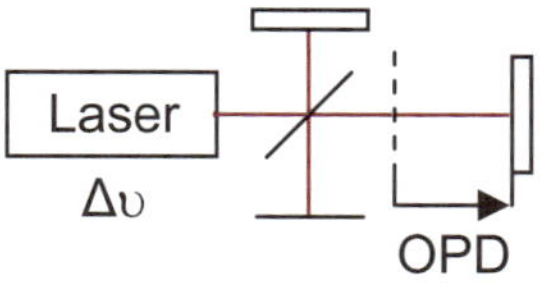

One drawback is that the frequency shift required for a 90° phase shift is part dependent because the OPD depends on the test part location.

Phase shifting methods involving a frequency difference between the test and reference beams are useful in **distance measuring interferometers**, or DMIs. A DMI often uses a **Zeeman laser**, which outputs two orthogonally polarized frequencies. Variations in the beat frequency between the two beams are translated into a displacement. Nanometer resolution is common.

In general, continuous phase shifting is useful for measuring something that is changing rapidly. A small $\Delta\nu$ can result in the ability to measure phase very rapidly so that phase evolutions in time can be recorded.

Phase Shifting Algorithms

Once the ability to introduce a phase shift is achieved, the gathered set of interferograms must be analyzed to find the phase at each pixel in the image, $\phi(x,y)$. The most common phase shift between images is $\pi/2$ (90°):

$$
\begin{aligned}
I(x,y) &= I_{dc}(x,y) + I_{ac}(x,y)\cos[\phi(x,y) + \phi(t)] \\
I_1(x,y) &= I_{dc} + I_{ac}\cos[\phi(x,y)] \qquad \phi(t) = 0(0^\circ) \\
I_2(x,y) &= I_{dc} - I_{ac}\sin[\phi(x,y)] \qquad \phi(t) = \pi/2(90^\circ) \\
I_3(x,y) &= I_{dc} - I_{ac}\cos[\phi(x,y)] \qquad \phi(t) = \pi(180^\circ) \\
I_4(x,y) &= I_{dc} + I_{ac}\sin[\phi(x,y)] \qquad \phi(t) = 3\pi/2(270^\circ)
\end{aligned}
$$

Some common algorithms for calculating the phase:

$$\text{Three Step: } \phi = \tan^{-1}\left(\frac{I_3 - I_2}{I_1 - I_2}\right)$$

$$\text{Four Step: } \phi = \tan^{-1}\left(\frac{I_4 - I_2}{I_1 - I_3}\right)$$

$$\text{Schwider-Hariharan (5 step): } \phi = \tan^{-1}\left[\frac{2(I_2 - I_4)}{2I_3 - I_5 - I_1}\right]$$

The minimum number of interferograms needed to determine $\phi(x,y)$ is three, but a fourth image helps reduce the error due to incorrect phase shifts. Further improvement occurs when a fifth image is taken, where $\phi(t) = 360^\circ$. This fifth image is nominally identical to the first, but this algorithm, known as the **Schwider-Hariharan** algorithm, is much less sensitive to miscalibration of the phase shifter. If the phase shift is not exactly 90° between images, errors show up in the phase at twice the frequency of the interference fringes. The fifth image greatly reduces this type of noise in the calculated phase. More images can be added for additional improvement in errors due to incorrect phase shifts. The calculated phase can be converted to OPD, which is then converted to part error depending on the interferometer setup. For a Twyman-Green interferometer testing a mirror, the height $h = OPD/2$.

$$OPD(x,y) = \frac{\lambda}{2\pi}\phi(x,y)$$

Basic Phase Unwrapping

The inverse tangent operation common to these phase shift algorithms unwraps the phase to the range $0 < \phi < 2\pi$. Often the phase across the images will need to be further unwrapped after modulo 2π correction. Upon inspection of the phase data, there will be phase discontinuities of approximately 2π. To remove these phase ambiguities, the phase at adjacent pixels must be less than π difference, or 1/2 a wave of OPD. Trace the path of the phase across the image and when the phase jumps by more than π, add or subtract $N2\pi$ until the phase difference is brought back to be less than π. The other requirement is that the phase is continuous, which is true of most optics. Optics with steps or large surface deviation will be discussed later.

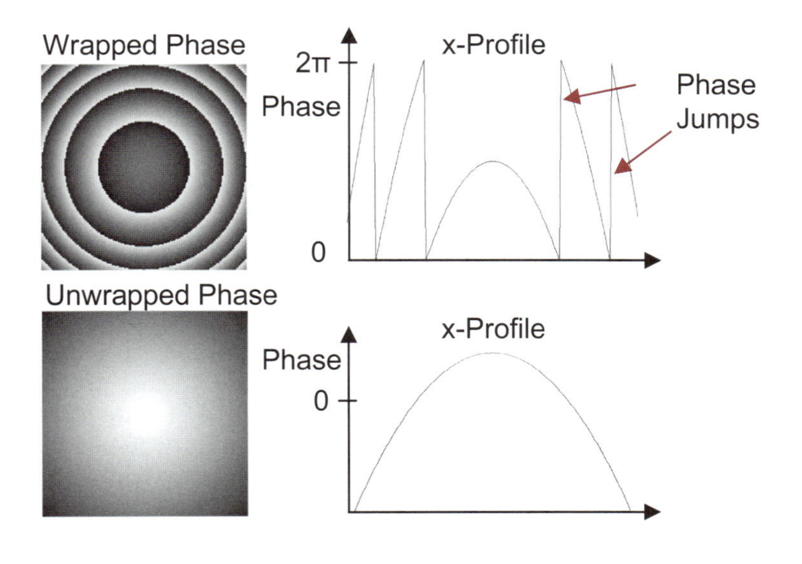

> Phase unwrapping in the presence of noisy regions in the images is the subject of many papers and books.

Phase-Stepping vs. Phase-Ramping

It is important to distinguish between **phase-stepping** and **phase-ramping** (integrated bucket). Phase-stepping is when the phase is changed, an interferogram is recorded, and the phase is changed again. Phase-ramping is when the phase is changed at a constant rate and the interferograms are taken at the appropriate time interval.

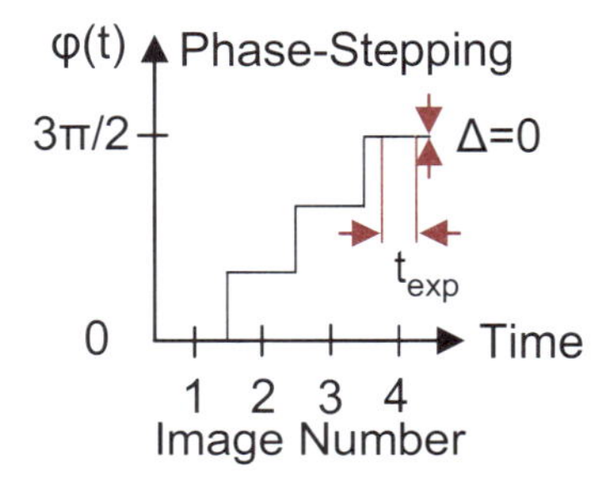

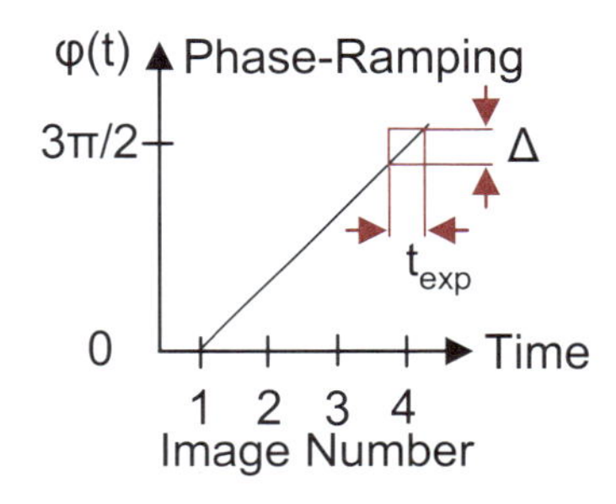

Phase-stepping takes longer since it is necessary to wait after stepping for the effects of ringing in the phase stepper to die out. One of the biggest sources of error in phase shifting interferometry is vibration in the system; therefore, the faster a measurement can be made, the better. Phase-ramping is faster since there is no need to wait for the system to settle, and the only trade-off is a slight decrease in fringe contrast. Since the exposure time (t_{exp}) for each image is finite, there is a phase change Δ over which the image integrates. Each image is the result of integrating over a phase change Δ, which results in a decrease in fringe contrast, given below. Γ is the initial fringe contrast.

$$I_i = I_0\left\{1 + \Gamma\,\mathrm{sinc}\left(\frac{\Delta}{2}\right)\cos[\phi(x,y) + \phi_i]\right\};\quad \mathrm{sinc}(x) = \frac{\sin(x)}{x}$$

For a Δ of $\pi/2$, there is only a 10% reduction in fringe contrast. The same algorithms for calculating the phase used for phase-stepping can also be used for phase-ramping.

Errors in PSI

For any combination of phase-shifting technique and associated algorithm for calculating $\phi(x,y)$, errors will be present in the data. Some of the most common error sources are stray reflections, quantization errors, detector nonlinearity, frequency and intensity instability in the light source, and incorrect phase shifts between data frames. Effects such as pixel-to-pixel nonuniformity and spatial intensity variations do not cause errors in the phase calculation since each pixel's phase is found independent of the neighboring pixels.

Stray Reflections

Lasers are great for interferometers in that their large coherence length allows optical separation of the test and reference arms. Unfortunately, stray reflections off the many surfaces in an interferometer will also interfere and create extraneous fringes. These extra fringes add irradiance and phase to the test beam and can cause large phase errors. The best way to eliminate these errors is to use a low-coherence light source or design the system such that stray reflections are blocked.

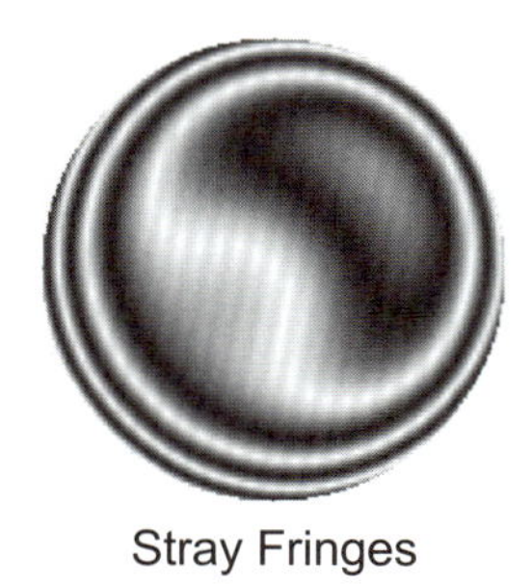

Stray Fringes

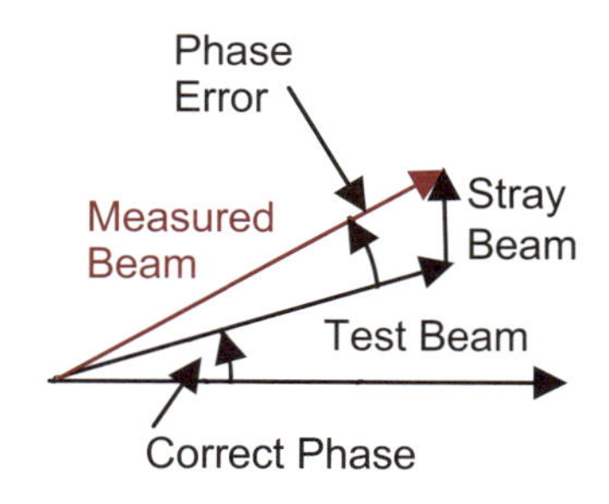

Quantization Errors

Interferograms are of course analog signals, but in order to process them in the computer, they must be digitized. Quantizing the signal will cause a phase error which decreases as more bits (levels) are used. The number of bits in this equation, b, is the bit depth used by the fringe modulation, not of the camera. $\sigma_{\phi,q}$ is the RMS phase error and N is the number of steps in the algorithm.

$$\sigma_{\phi,q} = \frac{2}{2^b\sqrt{3N}}$$

Quantization error can be reduced by averaging data sets if the noise in the signal is greater than one bit.

Detector Nonlinearity

Nonlinearity in a detector can cause phase errors in a measurement and care should be taken to adjust exposure time so as not to operate near saturation or at extremely low signal levels. Most detectors are extremely linear over most of their dynamic range, so this is not usually a large source of error in PSI.

Source Instabilities

Frequency instability in the light source will introduce a phase shift if the paths are not matched, i.e., $d \neq 0$. If $\Delta\nu$ is the frequency change in the source, $\Delta\phi_{Freq}$ is the phase error.

$$\Delta\phi_{Freq} = 2\pi\frac{d}{c}\Delta\nu$$

Irradiance fluctuations in the light source also introduce errors in the measured phase. If N is the number of steps in the algorithm and SNR is the signal to noise ratio, then the standard deviation of the phase is $\sigma_{\phi,I}$.

$$\sigma_{\phi,I} = \frac{1}{SNR\sqrt{N}}$$

Incorrect Phase Shift

Incorrect phase shifts between data frames come in two forms. The first is if the phase shifting mechanism is miscalibrated, so that the error is a linear error, ε.

$$i\frac{\pi}{2} \rightarrow i\frac{\pi}{2} + i\varepsilon \quad i = \text{Step Number}$$

A phase shift error between steps results in a phase error that is sinusoidal with a frequency of twice the frequency of the interference fringes. In general, adding more steps reduces the phase error. If the phase shift error is only due to a 5% miscalibration, going from 5 to 7 steps reduces the error by 3 orders of magnitude. Checking for this linear phase step error is done by calculating the average phase shift between frames. For a set of five interferograms, the average phase shift between frames, a, can be calculated for each pixel.

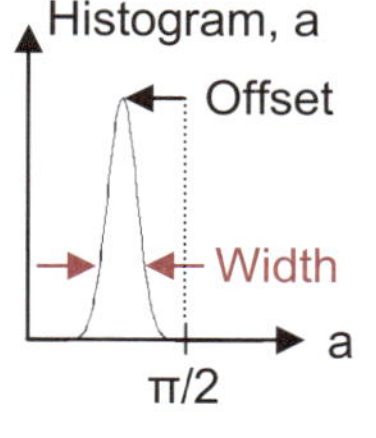

$$a(x,y) = \cos^{-1}\left[\frac{1}{2}\frac{I_5(x,y) - I_1(x,y)}{I_4(x,y) - I_2(x,y)}\right]$$

This algorithm often has singularities and tilt fringes can be introduced to help avoid them. Bad data points can be removed using a threshold and the values for a can be displayed in a histogram. Not only is this useful for adjusting the phase shifting interferometer such that the center of this histogram is at the correct value, but the width of the histogram indicates if there are other errors in the system, such as vibration.

Vibrations

The second cause of incorrect phase shifts is vibration in the interferometer. Vibrations are the most serious impediment to wider use of phase-shifting interferometry. Since PSI takes multiple images over a finite time period, it is sensitive to the time-dependent phase shifts due to vibrations. These vibrations are difficult to correct since the optimum algorithm depends on the frequency and the phase of the vibration. For a given vibration amplitude, the phase error is a function of the ratio of the vibration frequency relative to the frame capture rate. Vibrations at half the frame capture rate cause the largest phase error.

Avoiding Vibrations

Several methods have been used to try to get around the vibration problem in PSI. Speeding up the data collection process by phase ramping helps, but there are limits in the electronics for how fast data can be captured. The environment can be controlled, but this gets expensive. Common path interferometers are difficult to phase shift. All frames can be taken simultaneously, but this type of system can be difficult to align and calibrate. Some of these systems are described here.

Single-Shot Interferometer

One method for a single-shot phase-shifting interferometer is to have orthogonal polarization states in the test and reference beams of a Twyman-Green interferometer. After the beams are recombined, a **holographic optical element** (HOE) is used to create four identical copies of the beam that fall on a single CCD. Each of the four images passes through a different birefringent phase mask, which introduces a phase shift between the test and reference beams of 0°, 90°, 180° and 270°. Finally, a polarizer with its transmission axis at 45° to the two polarization states is placed after the phase mask so that the beams interfere.

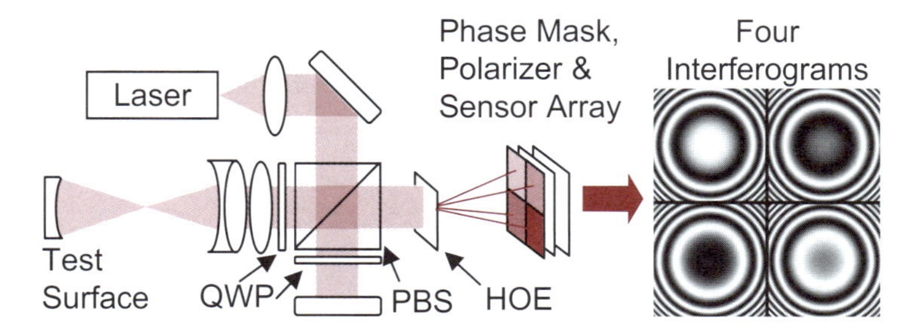

This system is compact and easy to align. By simultaneously capturing all four interferograms, this system is insensitive to vibration. By capturing several frames, the phase as a function of time can be measured and displayed in the form of a "phase movie".

Spatial Synchronous and Fourier Methods

Both of the following techniques require a single interferogram with a large amount of tilt. If the tilt between the two beams is in the x direction, the signal is given as follows:

$$I(x,y) = I_{dc}(x,y) + I_{ac}(x,y)\cos[\phi(x,y) + 2\pi fx]$$

$$\text{Ref1}(x,y) = \cos(2\pi fx) \quad \text{Ref2}(x,y) = \sin(2\pi fx)$$

$$\phi(x,y) = \tan^{-1}[-Sm(I * \text{Ref1})/Sm(I * \text{Ref2})]$$

The spatial frequency of the tilt fringes is f. This signal is multiplied by two reference signals. The high-frequency part is filtered out of each product using a smoothing function (Sm[ ]), and the phase can then be calculated.

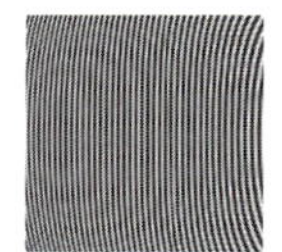
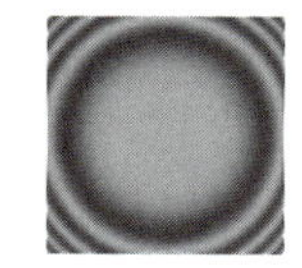
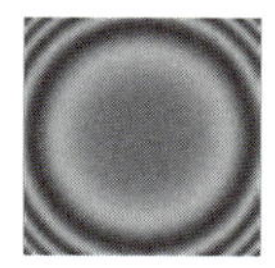
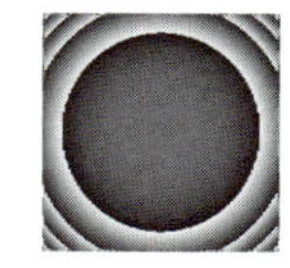

$I(x,y)$ $\quad Sm[I * \text{Ref1}]$ $\quad Sm[I * \text{Ref2}]$ $\quad \phi(x,y)$

Alternatively, the same interferogram $I(x,y)$ can be **Fourier transformed** (FT). It is then spatially filtered to select the portion of the Fourier transform around the spatial frequency f. An inverse FT of this filtered signal gives the wavefront phase.

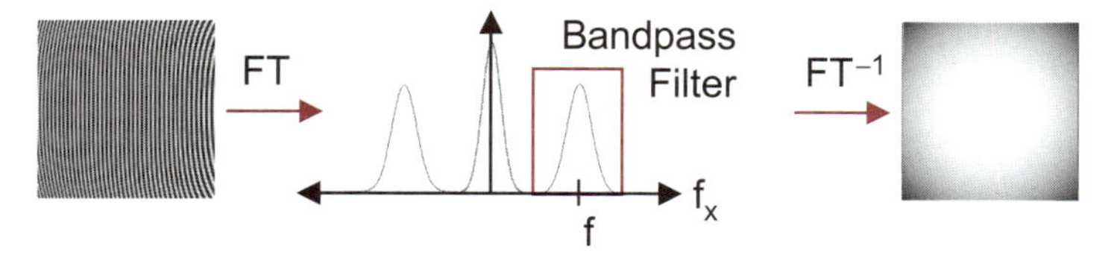

Both the spatial synchronous and Fourier method require a large amount of tilt to separate the orders. These tilt fringes could be introduced simply by tilting the reference mirror, but this will introduce **retrace errors**. A better method is to have orthogonal polarization states for the two beams and to introduce tilt using a Wollaston prism followed by a polarizer just before the detector. This large tilt limits the accuracy capabilities of these two methods. The advantage is that only a single interferogram is needed, avoiding issues with vibration.

Spatial Carrier Interferometry

Spatial carrier interferometry is similar to spatial synchronous in that a single interferogram is taken with a lot of tilt fringes, but now the assumption is made that the wavefront is nearly flat. For example, if the tilt is such that there are four detector elements between fringes, then there is a phase change of 90° between pixels. Normal phase-shifting algorithms can then be used to find the phase using the intensities of the four adjacent pixels.

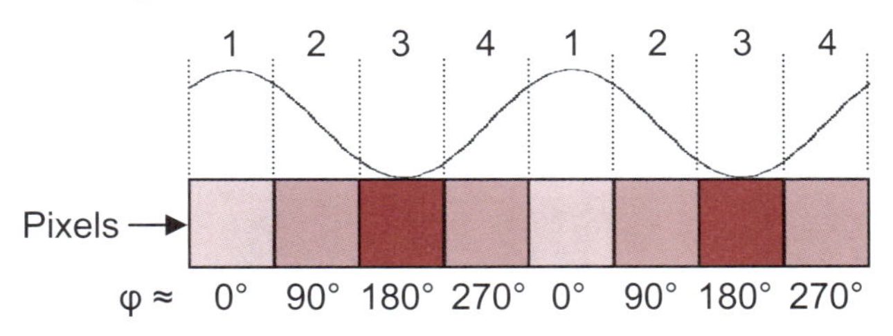

Micropolarizer Array, Phase-Shifting Interferometer

This method for instantaneous PSI requires the two beams to be orthogonally circularly polarized so that a linear polarizer at an angle α will create a phase shift of 2α. If the wavefront is relatively flat, 'phase pixels' can be made up of 2×2 pixel subsets of the detector. A polarizer mask can be placed in front of the camera to create the appropriate phase filter. Each group of four pixels is used to calculate ϕ for that 'phase pixel' using normal phase-shifting algorithms.

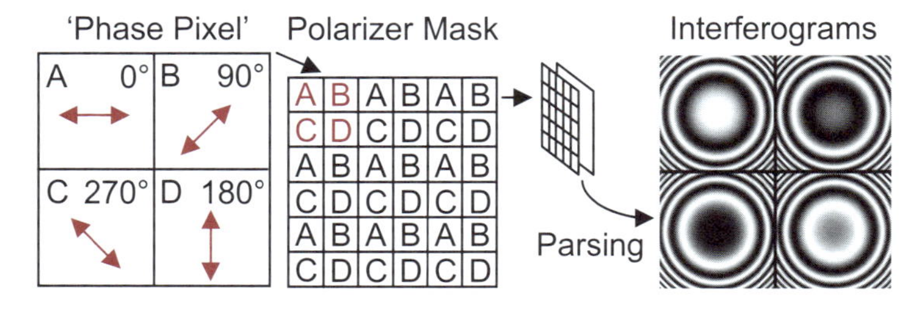

It should be noted that the above techniques are susceptible to detector nonuniformity errors, although modern cameras are usually quite uniform.

Ground Glass

Using a **ground glass diffuser** in an interferometer is useful for destroying spatial coherence. Ground glass is often used to limit the spatial coherence of the source. Coherence is a requirement to obtain interference fringes, but spurious fringes due to stray reflections are a dominant noise source. A laser is focused onto a rotating ground glass diffuser to decrease the spatial coherence and render stray reflections incoherent with the test and reference beams. A stationary diffuser creates a stationary speckle pattern, so the ground glass must be rotated so the speckle pattern changes much faster than the camera integration cycle. Each scatter site on the ground glass has a random phase. Integrated over time, the phase distribution at each location becomes uniform.

In order to increase the flexibility of commercial laser-based Fizeau interferometers, a zoom lens can be used to adjust for varying test part sizes. A multi-element zoom lens creates many stray reflections, which cause spurious fringes in the recorded interferogram. Imaging the interferogram onto ground glass before the zoom lens converts the two coherent waves into an incoherent irradiance signal that is imaged to the camera via the zoom lens. Any stray reflections within the zoom lens are incoherent; they add in irradiance and do not cause phase errors. Ground glass scatters light, causing a large amount of loss in the system. A second drawback is that the motor that rotates the ground glass inevitably introduces vibrations, another major noise source in interferometers.

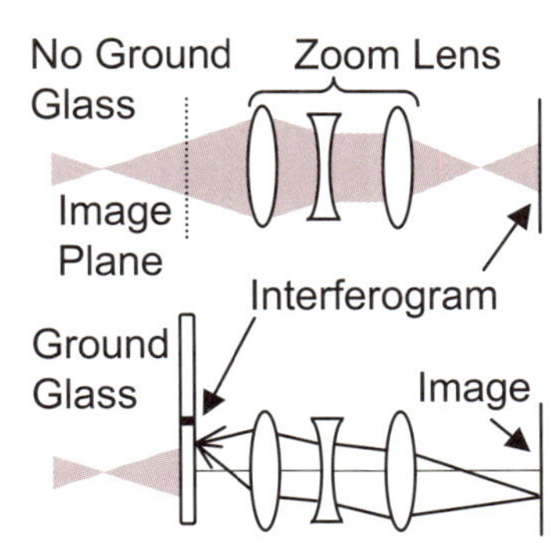

Surface Microstructure

Interferometry can also be applied to characterization of surface quality. **Surface quality** refers to the finish of an optical surface, including pits, scratches, incomplete polish or stain. In many applications, surface quality will not affect the overall performance of an optical system, but there are exceptions, such as if that surface is near a focal plane, if the system is especially sensitive to stray radiation, or if high-energy lasers may cause damage to the part due to surface features. Optics are usually accompanied by a **scratch-dig** specification, ranging from 80-50 to 10-5. The first number is the apparent width of the largest scratches on the surface in microns. The second number is the largest diameter of a permissible dig, pit or bubble in hundredths of a millimeter (50 is 0.5 mm diameter). The methods discussed here are those based on interferometry; other methods are used, such as mechanical probes or atomic force microscopy.

Nomarski Interference Microscope

The **Nomarski microscope** measures surface slope changes and is also referred to as a **differential interference contrast microscope** (DIC) or a **polarization interference contrast microscope**. A white light source is used, followed by a polarizer which sets the angle of the polarized light incident on a Wollaston prism. The Wollaston is made of two uniaxial crystals (different index of refraction for the two polarization states) with optical axes as shown.

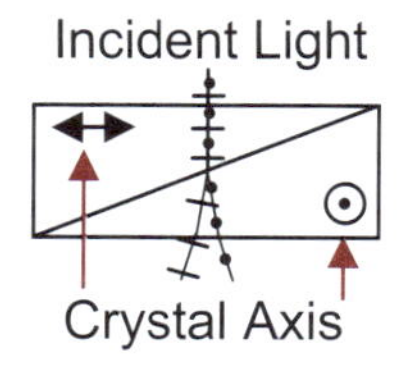

The change in index causes a polarization-dependent refraction at the crystal interface. The Wollaston prism splits the light into two beams of orthogonal polarizations propagating at slightly different angles.

Nomarski Interference Microscope

The two sheared beams are both incident on the test surface but are slightly offset as a result of the shear. The shear at the test surface is set to be comparable to the optical resolution of the microscope objective, so only one image will be seen in the image plane. The beams reflect back through the prism and are recombined. In order to have interference, an analyzer (polarizer) at a fixed angle is placed before the camera. The resulting interferogram is a map of the slope changes of the surface. The slope information is only obtained in the direction of the shear and the part must be rotated to obtain the other direction. Since a white light source is used, colors in the image indicate varying slopes.

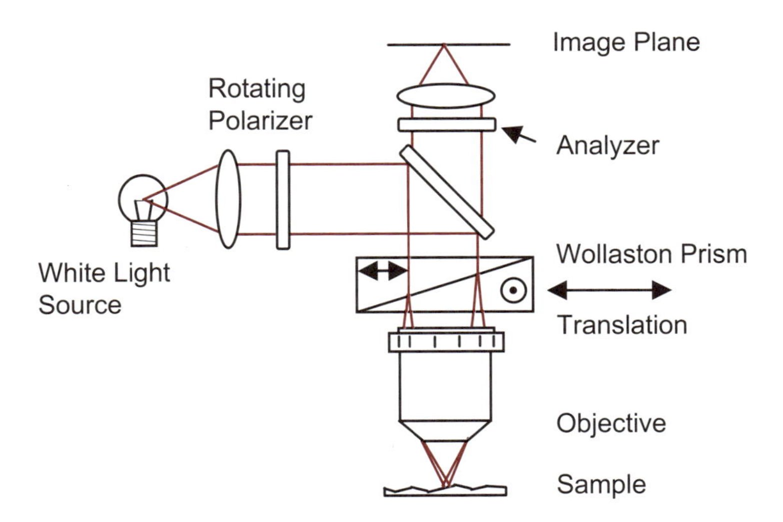

Rotating the first polarizer changes the relative intensities of the two beams, which can change the colors. When the Wollaston is centered and a perfect surface is examined, the OPD between the two beams is 0. Translating the Wollaston orthogonal to the beam introduces an OPD and changes the color. The color ultimately depends on the OPD between the two beams at each point.

Fringes of Equal Chromatic Order (FECO)

The **FECO** interferometer is a multiple-beam interferometer using a white-light source. The test surface is placed next to the reference surface to form a **Fabry-Perot cavity**, where the separation is d, the index in between is n, and the beam angle in the cavity is θ. The transmitted irradiance I_t depends on the coefficient of finesse, F. Here, ρ is the ratio of reflected irradiance and ϕ_r is the phase change on reflection at each surface.

$$I_t = \frac{I_{inc}}{1 + F\sin(\phi/2)}$$

$$\phi = \frac{2\pi}{\lambda_{air}} 2nd\cos(\theta) + 2\phi_r$$

$$F = \frac{4\rho}{(1-\rho)^2}$$

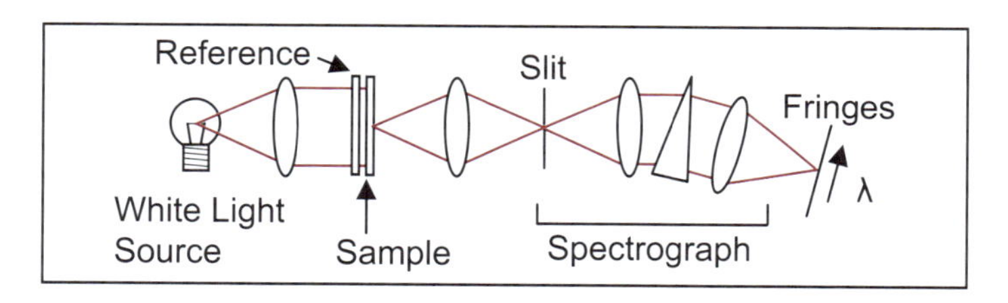

Both the sample and reference need to have high reflectance so that the multiple beam interference fringes have high finesse. The other drawback is that data is only obtained along the line chosen by the slit entrance to the spectrograph, where the slit is open into the page, along the x-axis. Examining the fringes gives the separation variation of the two surfaces along x. If the reference is known to be flat, the separation variations can be translated into sample surface variations.

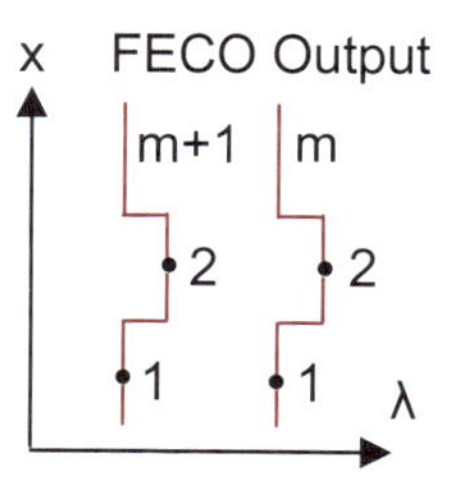

$$d_2 - d_1 = \frac{\lambda_{1,m+1}}{\lambda_{1,m} - \lambda_{1,m+1}} \left(\frac{\lambda_{2,m} - \lambda_{1,m}}{2} \right)$$

Phase-Shifting Interference Microscope

Interference microscopes provide a fast, noncontact method for mapping surface microstructure in 2D or 3D with excellent vertical resolution. Four or five frames are captured as in traditional PSI. The light source used is typically a spectrally filtered white light source, such that the bandwidth is around 40 to 80 nanometers for a coherence length of a few microns. This gives the advantage of eliminating extraneous fringes from other surfaces while giving a long enough L_c so that fairly large surface topography variations can be measured. The lateral resolution is limited by the optical resolution, as with any optical microscope.

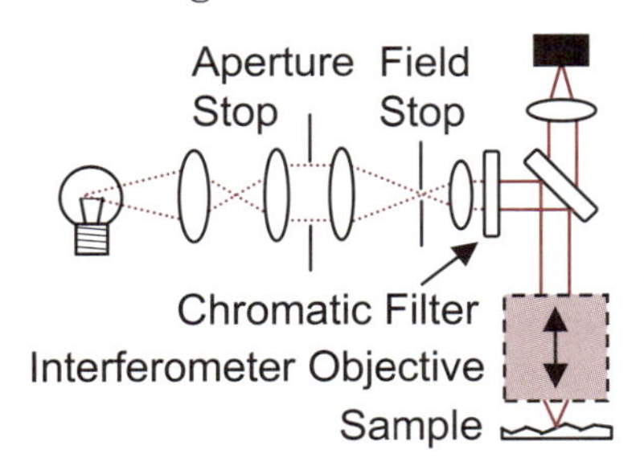

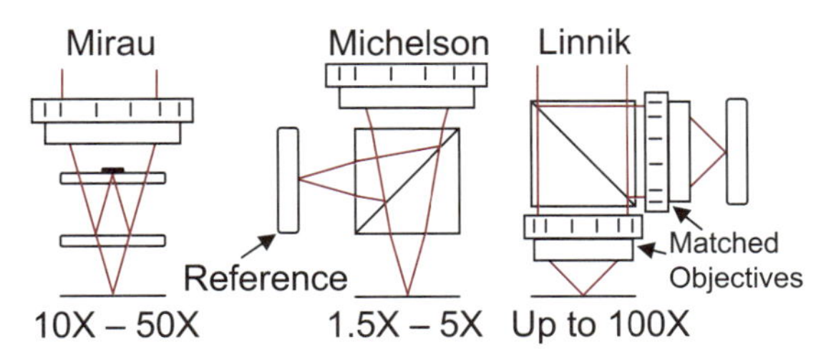

The **Mirau**, **Michelson**, and **Linnik** interferometer objectives are three of the different objectives used in a PSI microscope, and each is useful for different magnification ranges. The Mirau has a central obscuration due to the reference mirror, while the working distance of the Michelson is limited by the size of the beamsplitter. The Linnik is good for large magnifications, but requires two matched objectives, making it expensive. One important disadvantage of all interferometric microscopes is that any variation in the phase change on reflection due to different sample composition (metals vs. dielectrics) will be interpreted as a false height variation. The largest step height between adjacent pixels that can be measured unambiguously with a PSI microscope is $\lambda_c/4$, where λ_c is the center wavelength of the filter.

Multiple-Wavelength Interferometer

One method for overcoming the dynamic range limit in PSI is to use multiple wavelengths. Single-wavelength phase-shifting interferometry can be extremely accurate, but the dynamic range is limited to an OPD of $\lambda/2$, which is a step height of $\lambda/4$ for a reflection test. When measuring surface microstructure, such as a step, there is a discontinuity in the fringes. The fractional phase difference is easily found, but more information is required to know the fringe order, or how many full waves of OPD difference there are at the step.

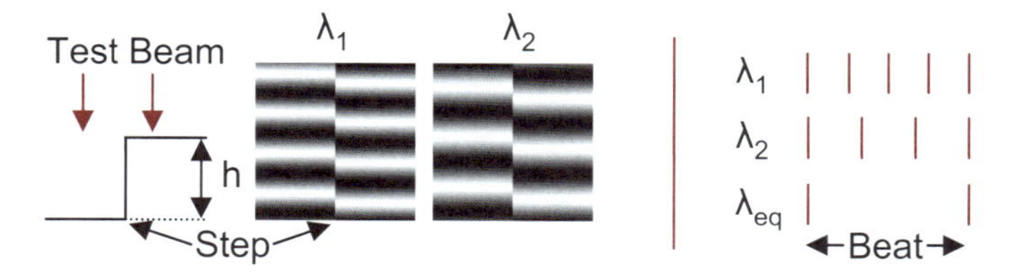

If the same measurement is repeated using a second wavelength, the dynamic range is increased to the equivalent wavelength, λ_{eq}. As long as the step height is less than $\lambda_{eq}/4$, there will not be any height ambiguities in the step height measurement. The method can also be thought of as measuring the step using the beat frequency. As λ_1 and λ_2 get closer together λ_{eq} increases, but this is only practical so long as the difference between the two phase changes across the step for the different wavelengths is greater than the uncertainty in the measured phase values. A better way to increase the dynamic range further is to add more wavelengths.

$$\lambda_{eq} = \frac{\lambda_1 \lambda_2}{|\lambda_1 - \lambda_2|} \qquad OPD = 2hn = \frac{\lambda_{eq}}{2\pi n}(\Delta\phi_{\lambda 1} - \Delta\phi_{\lambda 2})$$

Multiple-wavelength interferometry is difficult to use in transmission since the index varies with wavelength.

λ_1 (μm)	λ_2 (μm)	λ_{eq} (μm)
0.633	0.532	3.3
0.633	0.612	18.4
0.543	0.488	4.8
0.488	0.483	47.1

Vertical Scanning Techniques

Another way to increase the dynamic range of interferometry is to use **vertical scanning white light interferometry** (VSWLI). The apparatus for a vertical scanning, or coherence probe interferometer is similar to the PSI microscope, including the types of objectives. The full spectrum of the white light source is used to obtain a narrow visibility function, V(OPD). As the objective scans relative to the sample, the fringes are modulated by the visibility (coherence) function, which is maximized when the OPD between the test and reference arms is 0.

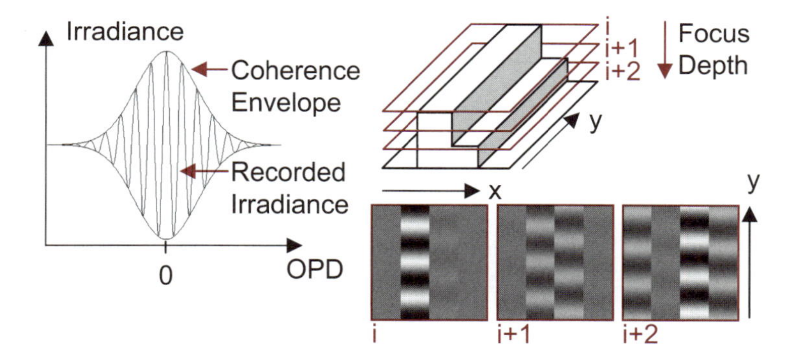

Demodulating the fringe signal gives the peak of the coherence function, which corresponds to the relative surface height at that spatial location. Since this technique uses the peak of the coherence envelope to find the surface, errors due to phase changes on reflection across the sample will be reduced. Unlike PSI, the dynamic range is not limited to $\lambda/4$ but is only limited by the travel range of the objective. This travel range is on the order of millimeters and depends on the working distance of the objective. This allows vertical scanning interferometry to measure surfaces from polished optics to print rollers to paper to a quarter, all with a resolution of a few nanometers.

Flat Surface Testing

Now that the basic theory of interferometry and the most common interferometer configurations have been covered, discussion on testing various optical components can begin. In all cases, an interferogram is created from the superposition of a test beam and a reference beam and interpreted to determine surface characteristics. It is important to keep in mind that the interferogram is the *difference* between the two beams. When measuring a test surface for flatness, the reference surface must be characterized absolutely so that the interferogram is known to be the deviation from flatness of the test part. Establishing an absolute flat will be covered later.

Mirrors

The easiest interferometric method for testing many flat surfaces is the classical Fizeau with an established flat as the reference surface. In the case of a mirror, the reference must be on top of the test surface and the fringe visibility will be about 0.4 ($\rho_{ref} \sim 0.04$, $\rho_{test} \sim 1$). Various surface errors and the resulting interferograms are shown here.

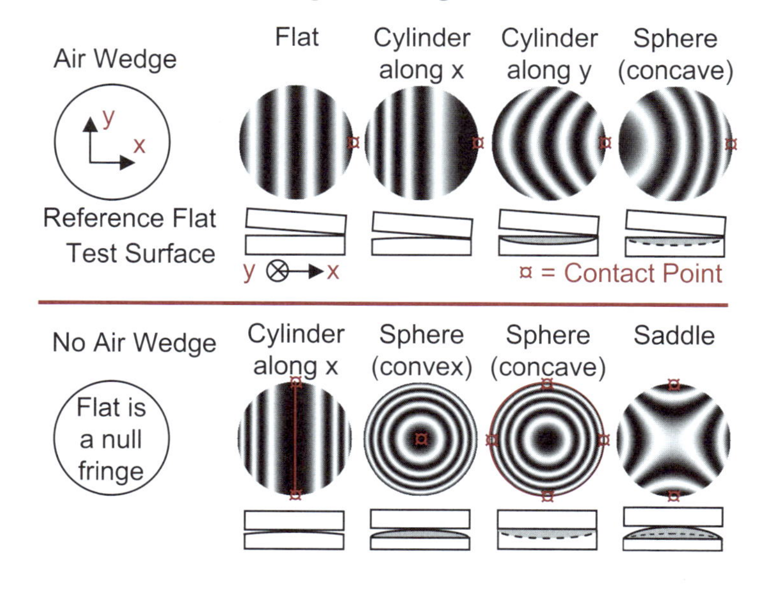

Mirrors—Continued

A more accurate method is a phase-shifting laser-based Fizeau interferometer. The interferograms would be the same as for the classical Fizeau with the accuracy gain due to the phase shifting and subsequent computer processing. Another advantage is that this is a non-contact test, which eliminates the risk of scratching the test surface. Another option is the Twyman-Green, which is more versatile, allowing the testing of a flat larger than the reference by using a beam expander in one arm of the interferometer. Again, the interferograms are the same. A disadvantage is that more high-quality optics are needed.

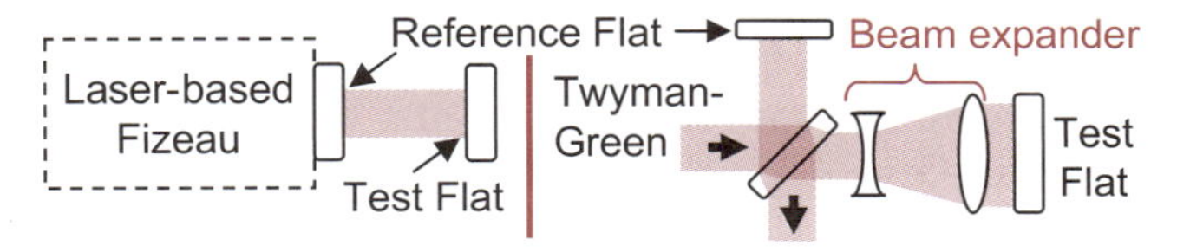

All of these techniques can be used for any flat optical surface test, such as the surfaces of a window or prism.

Windows

To complete the characterization of a window, the parallelism and thickness variation between the two surfaces must be measured. Both can be measured using a laser-based Fizeau or Twyman-Green without PSI.

Double Pass Transmission
Reference Flat
n
Test Window
t
D
Return Flat

$$OPD_{Measured} = 2(n-1)\delta t$$

$$OPD_{Wedge} = 2(n-1)\alpha D$$

$$\alpha = \frac{\lambda}{2(n-1)S}$$

This shows the setup for the Fizeau, where both the reference surface and return flat need to be characterized. Start by adjusting the return flat such that a single fringe covers the whole area. Then insert the window to be tested. Any fringes indicate a change in OPD due to thickness variations (Δt), with tilt fringes indicating a wedge (α, radians) between the two surfaces. The fringe spacing, S, gives the wedge angle. If the window is placed in one arm of a Mach-Zehnder interferometer, the above equations are the same except the 2's become 1's.

Windows—Continued

A second technique is to place the window in a collimated laser beam and look at the interference between the beams reflected off the two surfaces of the window.

$$\alpha = \frac{\lambda}{2nS}$$

If a phase-shifting interferometer is available, the above techniques will work. A simpler technique, which does not require a null fringe prior to sample insertion, is if the window is inserted in part of the beam of a laser-based Fizeau or Twyman-Green. Using PSI determines the direction of tilt in regions 1 and 2. The tilt difference $(\Delta\beta)$ between the two regions gives the window wedge.

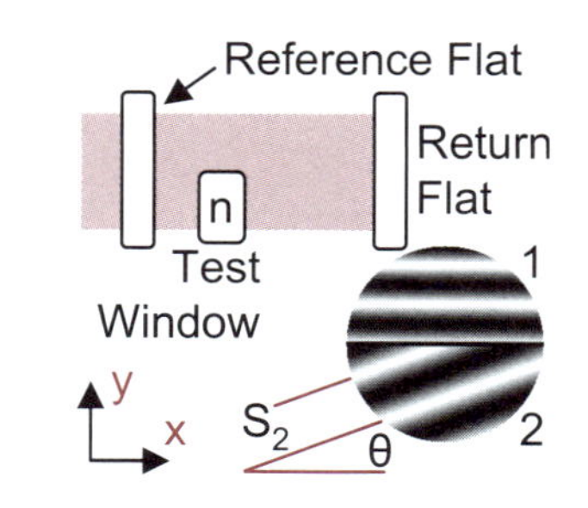

$$\alpha = \frac{\delta\beta}{2(n-1)} \qquad \beta_{x1} = \frac{\lambda}{S_{x1}} = \text{Tilt}$$

$$S_{x1} = \frac{S_1}{\sin(\theta)} \qquad S_{y2} = \frac{S_2}{\cos(\theta)}$$

$$\delta\beta = \sqrt{(\beta_{x1} - \beta_{x2})^2 + (\beta_{y1} - \beta_{y2})^2}$$

Prisms

Several parameters of a prism need to be tested: surface accuracy of prism faces, accuracy of angles, material homogeneity and transmitted wavefront accuracy. Surface accuracy can be measured using the methods described for mirrors. If a 90-degree prism is used to reverse a beam, the beam deviation θ is related to the error in the prism angle ε as follows:

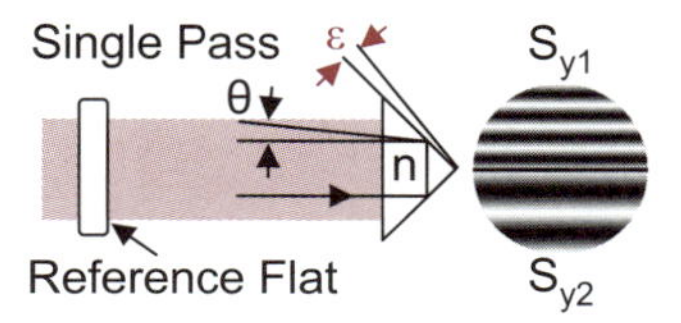

$$\varepsilon = \frac{\theta}{2n}, \quad \varepsilon = \frac{\delta\beta}{4n} = \frac{|\lambda/S_{y1} - \lambda/S_{y2}|}{4n}$$

The beam deviation θ can also be measured in a laser-based Fizeau or Twyman-Green in the single pass test. Any tilt difference $\Delta\beta$ between the two sides of the interferogram gives the prism angle error. The sign of the tilt is important in determining ε. Any errors in the collimated beam do not cancel in this setup. An improved method is the double pass test, where errors in the beam do cancel. The reference beam simply reflects off the reference flat while the test beam goes through the prism, reflects off the bottom of the reference and traverses the prism a second time. Only one interferogram is captured in this setup, and any tilt in the direction perpendicular to the seam of the prism indicates an error in the prism angle.

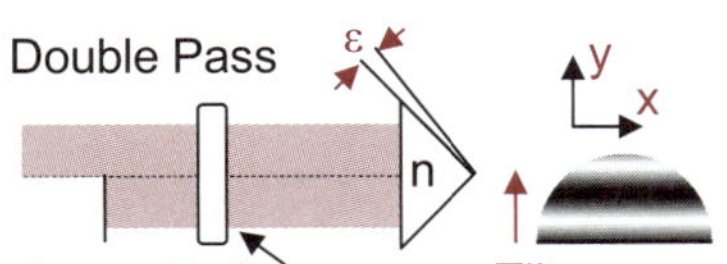

$$\varepsilon = \frac{\beta_y}{4n}$$

Transmitted wavefront accuracy is measured by placing a return mirror perpendicular to the beam transmitted through the prism and looking for OPD variations in the interferogram. These OPD variations are due to surface errors if indicated in the surface measurements or material inhomogeneity if not seen in the surface tests.

Reference Flat
Test Prism
Return Flat

$$OPD_{Measured} = 2(n-1)\delta t$$

Corner Cubes

Corner cubes are tested using the same methods for prisms already discussed. For the single pass test, the interferogram now has 6 sections. Straight fringes are obtained for a perfect corner cube. The tilt difference $\Delta\beta$ between two sections gives an angle error (ε) in the corner cube. For the double pass test, there will be no fringes from a perfect corner cube. Tilt fringes indicates that $\varepsilon \neq 0$.

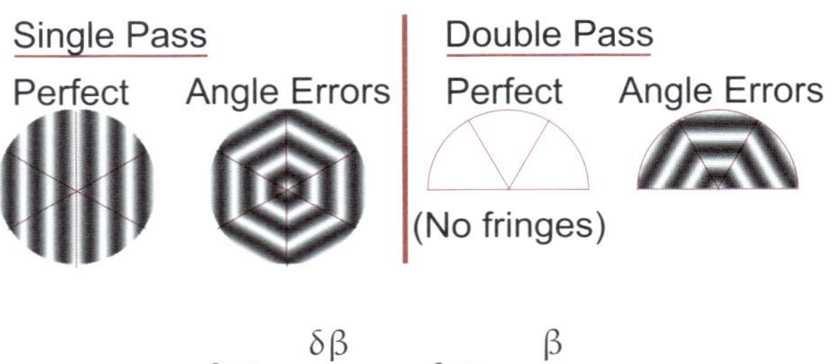

$$\varepsilon = \frac{\delta\beta}{3.266n} \qquad \varepsilon = \frac{\beta}{3.266n}$$

Diffraction Gratings

Reflection diffraction gratings can be tested using a Twyman-Green or Fizeau interferometer. Transmission gratings can be tested in a Mach-Zehnder.

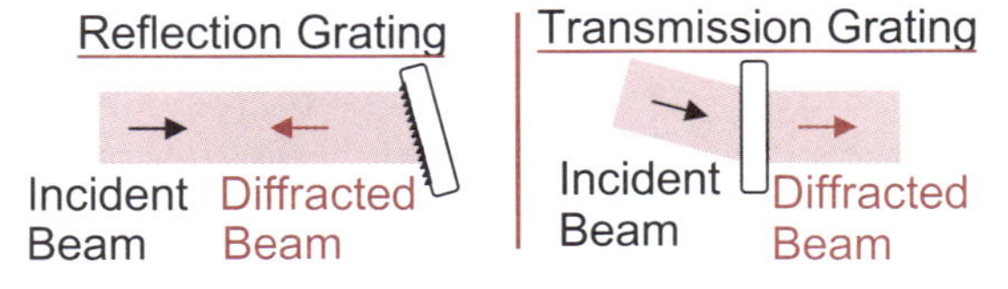

A straight line grating with unknown errors $\Delta(x, y)$ and line spacing Δx is represented as follows, where m is an integer. This is like an interferogram formed by a plane wave tilted at θ and a wavefront with aberration $W(x, y)$.

$$\frac{x}{\Delta x} + \frac{\delta(x, y)}{\Delta x} = m \qquad \frac{x \sin(\theta)}{\lambda} + \frac{W(x, y)}{\lambda} = m$$

$$\frac{W(x, y)}{\lambda} = \frac{\delta(x, y)}{\Delta x}$$

Any aberration measured in the first diffracted order is directly related to errors in the grating. For the Nth diffracted order, there will be N times the aberration.

Testing Curved Surfaces—Test Plate

For curved optics with large radii of curvature (ROC), the classical Fizeau can be used to determine R by placing the optic under test on top of a flat. The radius R is found by counting the number of bright fringes to a radial distance r_m, where the first bright fringes is $m = 0$.

$$R = \frac{r_m^2}{\lambda\left(m + \frac{1}{2}\right)}$$

The surface figure given by the shape of the fringes, as described earlier. Concave or convex surfaces with smaller radii can be tested by placing them on a known reference surface with about the same radius of curvature but opposite inflection. Here m is the number of fringe spacing the fringes depart from straightness, d is the test part diameter, and R is the radius of the reference part.

$$\Delta R = \frac{4m\lambda R^2}{d^2}$$

Aspheres can be tested this way, as long as the reference surface is well known so any fringe deviation can be attributed to the test part.

Twyman-Green Interferometer, ROC Test

When testing a spherical surface, there are two positions that give a null fringe. The first is sometimes called the **cat's eye** position, where the light comes to focus at the test surface. The second is the normal test setup for measuring surface figure, where the center of curvature of the test surface is coincident with the focus of the diverger lens. The axial distance the part is moved between these two positions is the radius of curvature.

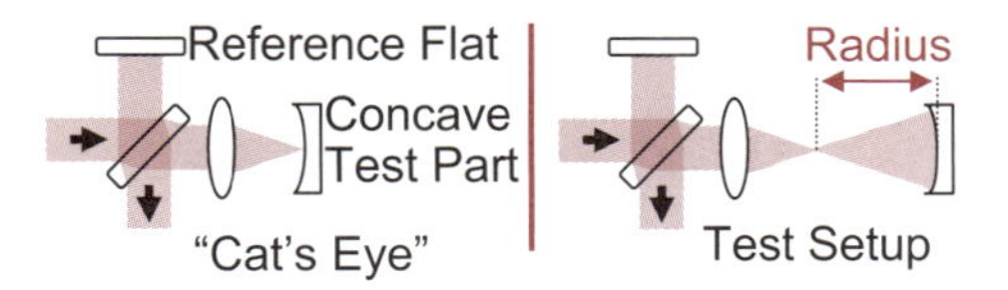

This technique also works with a laser-based Fizeau.

Curved Surfaces—Twyman-Green

The advantage of either the Fizeau or Twyman-Green setup for testing concave parts is that any size part can be tested as long as a fast enough beam (small $f/\#$) can be created using the diverger lens and the coherence length of the source is long enough. The $f/\#$ of the beam must be less than or equal to the $f/\#$ of the surface under test.

$$f/\#_{Beam} = \frac{f_{Div}}{D_{Div}} \approx \frac{1}{2NA}$$

$$R/\#_{Surface} = \frac{R}{D_S}$$

$$f/\#_{Beam} \leq R/\#_{Surface}$$

The Twyman-Green can also be used to measure convex surfaces with the additional restriction that the diameter of the test surface must be less than the diameter of the test beam.

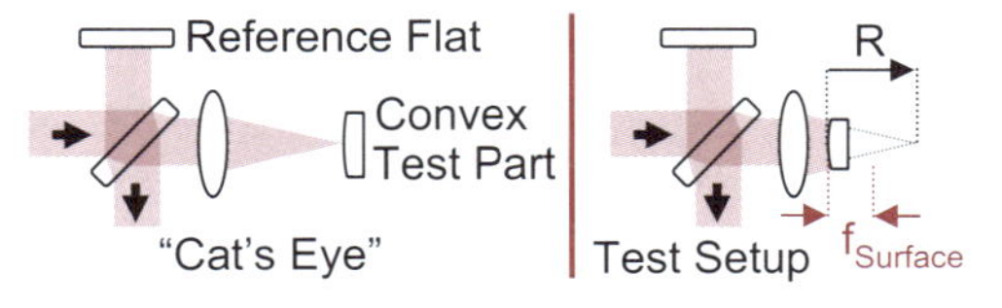

To measure the figure of a test surface, the light in the test beam should be normally incident on the surface. A null fringe indicates the entire test beam is normally incident. Incorrect spacing between the diverger lens and the test surface will introduce defocus fringes in the interferogram. In a phase-shifting interferometer, the test part should be adjusted to minimize the number of fringes. Typically, tilt and defocus are subtracted from the measured wavefront to give the surface figure.

$$1 \text{ Fringe} = 1\lambda \text{ of OPD} = \frac{\lambda}{2} \text{ Surface Height Deviation}$$

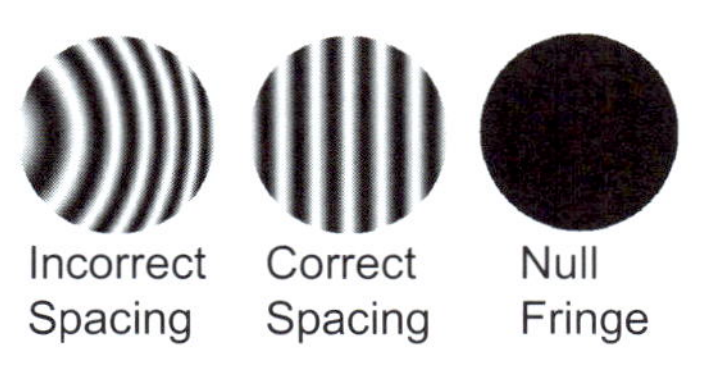

Curved Surfaces—Laser-based Fizeau

With care, the Twyman-Green can provide high-accuracy measurements; however, the test is expensive and subject to vibration and turbulence. Vibrations can be minimized by mechanically coupling the interferometer and the test part, giving a more stable interferogram.

Laser-based Fizeau

Curved surfaces are tested in a laser-based Fizeau in the same manner as in the Twyman-Green. Since the reference surface is now that last surface of the diverger, the required quality of the diverger optics is not as high.

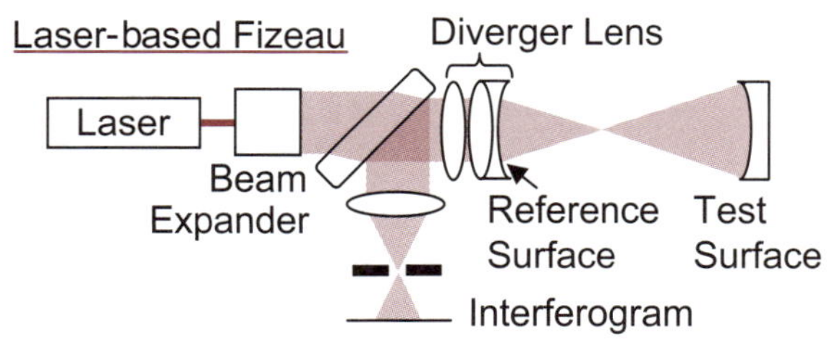

It is also easier to mechanically couple the reference and test surfaces to reduce the effects of vibrations. It is not practical to vary the reflectance of the reference surface to match the test part or to change the OPD for a given test part since the reference is fixed. Many commercial interferometers are of the Fizeau configuration since the diverger optics are less expensive and they are easily interchangeable to accommodate a variety of test parts.

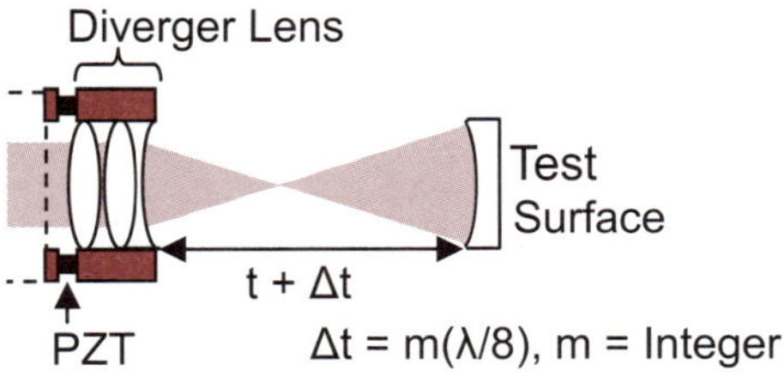

A laser-based Fizeau is commonly phase-shifted using a PZT mounted on the diverger lens assembly. It is shifted in $\lambda/8$ steps, changing the distance between the test and reference surfaces. The steps are small enough such that the lateral magnification change is negligible.

Testing Lenses or Lens Systems

Both the Twyman-Green and laser-based Fizeau interferometers can be used to test lenses in transmission. A positive lens system with a small enough aperture can be tested as shown here, where a high-quality convex or concave sphere with an $f/\#$ less than the lens system is used to retroreflect the test beam.

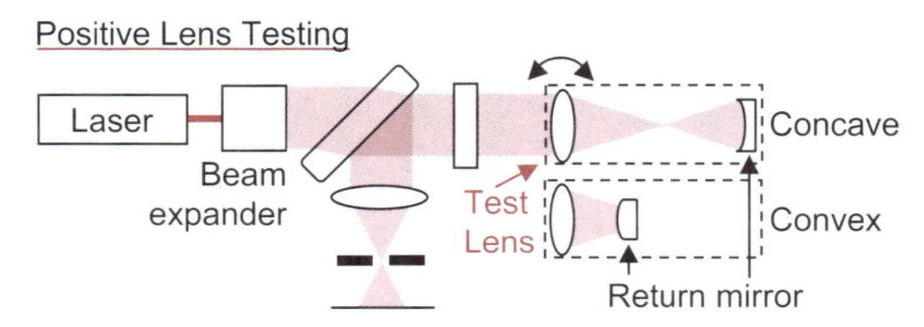

As shown below, larger diameter positive lenses can be tested. The $f/\#$ of the test lens must be greater then the $f/\#$ of the diverger lens. This setup is more difficult to align. In either setup, the test lens can be rotated to measure off-axis lens performance.

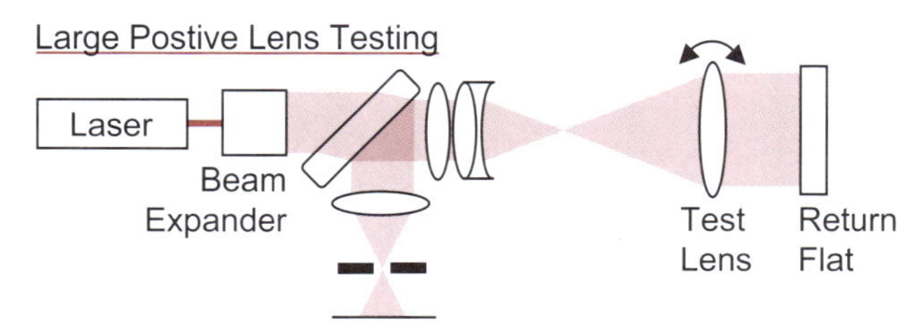

Negative lenses can be tested using similar methods and a large enough concave return mirror. The $|f/\#|$ of the test lens must be greater than or equal to the diverger or concave return mirror.

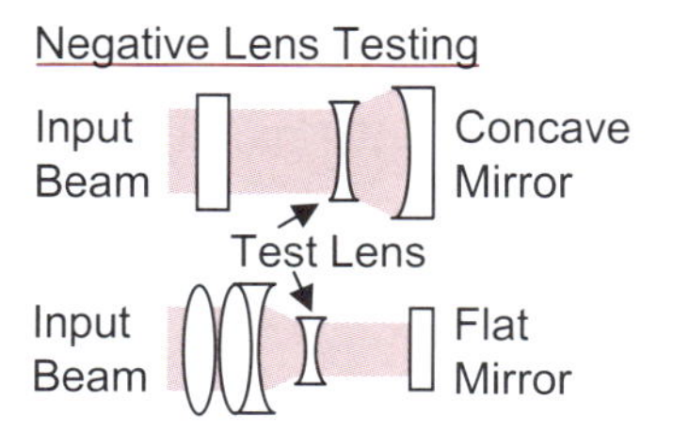

Shack Cube Interferometer

The **Shack cube interferometer** is a modification of the laser-based Fizeau that can be used to test the surface figure of concave surfaces. The advantage of this interferometer is that only the curved surface on the Shack cube beamsplitter needs to be of high quality since every other optic is

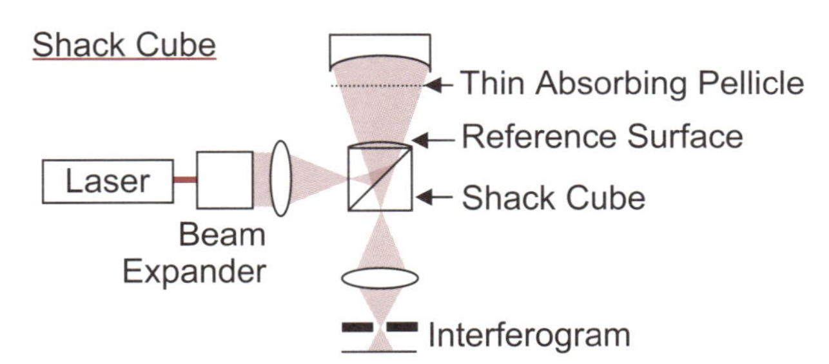

common path. In order to test lenses or convex surfaces, additional high-quality optics are required. This interferometer is lower cost than a Twyman-Green or regular Fizeau, but not as flexible. A pellicle can be used to help match the beam intensities.

The **scatterplate interferometer** is used for testing concave mirrors and gives interferograms that are interpreted the same way as those from a Twyman-Green or laser-based Fizeau. The advantage of this setup is that it is a common path interferometer, so the auxiliary optics' quality is unimportant. A laser can be used as the source, but it is often better to use a spectrally filtered white light source or arc lamp. The spatially filtered light source is imaged onto the test part and the scatterplate is placed near the center of curvature of the test surface so that the scatterplate is reimaged on itself (inverted).

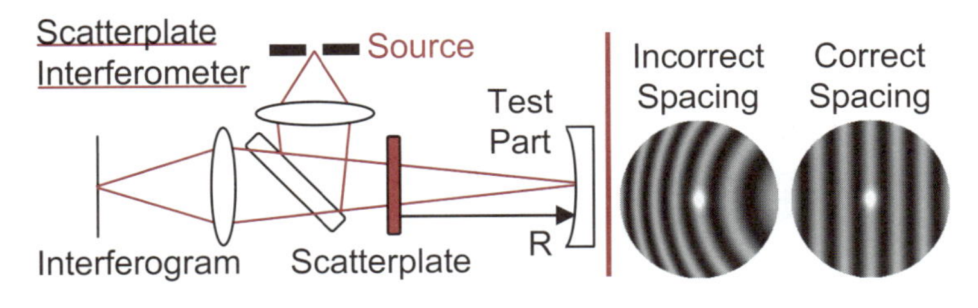

Scatterplate Interferometer

The light transmitted by the scatterplate the second time falls into four categories: (1) unscattered-unscattered, which creates a bright spot; (2) unscattered-scattered, the reference beam; (3) scattered-unscattered, the test beam; and (4) scattered-scattered, a background signal that reduces fringe contrast.

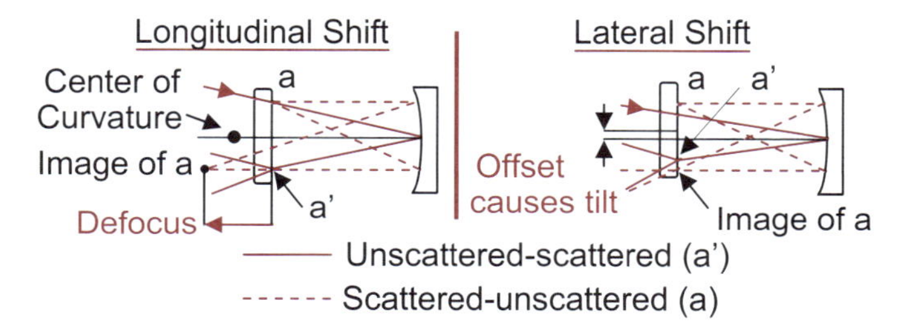

Longitudinal translation of the scatterplate changes defocus and lateral displacement introduces tilt.

The critical component of this interferometer is the scatterplate, which needs inversion symmetry. This can be done by exposing a photographic plate to laser speckle from a laser incident on a piece of ground glass, rotating the plate 180°, and superimposing a second exposure of the same speckle. To ensure uniform illumination of the test part from the scattered light, the solid angle of the ground glass seen from the photographic plate during construction of the scatterplate should be at least as large as the solid angle of the test part seen from the scatterplate. The plate should scatter 10 to 20% of the incident light.

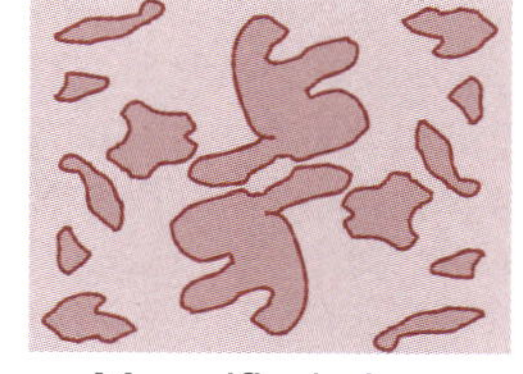

Magnified view of Scatterplate

$$\Omega_c \geq \Omega_t$$

Construction
Ground Glass
θ_c
Ω_c
Scatterplate
Interferometer
Ω_t
θ_t
Test Part

Phase-Shifting Scatterplate Interferometer

The difficulty in improving measurement accuracy by phase shifting with this interferometer is that the test and reference beams traverse nearly the same path, so shifting the scatterplate will not create a path difference. One way to phase shift utilizes a **birefringent scatterplate**. The scatterplate can be made of calcite, where the side with the scatter pattern and a flat piece of glass create a cavity containing index-matching oil that matches the index of the ordinary index of the calcite at the source wavelength. A spectrally narrow source (laser) is used so the chromatic variations of the polarization elements have a minimal effect. The first polarizer passes linearly polarized light at 45° with respect to the optic axis of the birefringent scatterplate, so half the beam will see the ordinary index of the crystal and not scatter because of the index-matching oil. The other half of the beam will partially scatter due to the index mismatch between the extraordinary index and the oil. The quarter-wave plate at 45° exchanges the two beams so that the unscattered reference beam scatters in the birefringent scatterplate, while the scattered test beam is unscattered on the second pass through the scatterplate. Note that if there is no unwanted scattering of the direct beams there will be no background irradiance term. If all the light scatters for the scattered beam, the hotspot is eliminated.

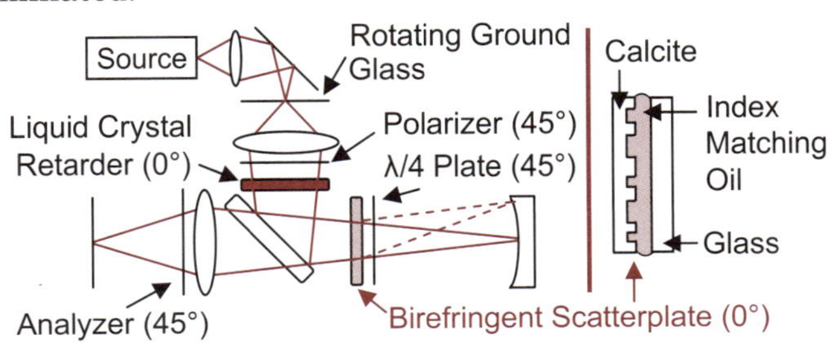

An analyzer at 45° is needed to combine the beams so they interfere. A liquid crystal retarder at 0° before the beamsplitter introduces a variable phase shift between the two beams. This setup performs comparably to a commercial Fizeau interferometer and is limited by the accuracy of the liquid crystal retarder.

Long-Wavelength Interferometry

During the fabrication of an optical surface, it is important to have a method for monitoring the surface figure during grinding. In the early fabrication stages, the surface may be too rough for visible wavelength testing. **Long-wavelength interferometry** provides a reduced sensitivity test for monitoring the surface figure of rough surfaces during fabrication. For a rough surface with a Gaussian height distribution with standard deviation σ, the fringe visibility reduction (V) is given as:

$$V = \exp\left(\frac{-8\pi^2\sigma^2}{\lambda^2}\right)$$

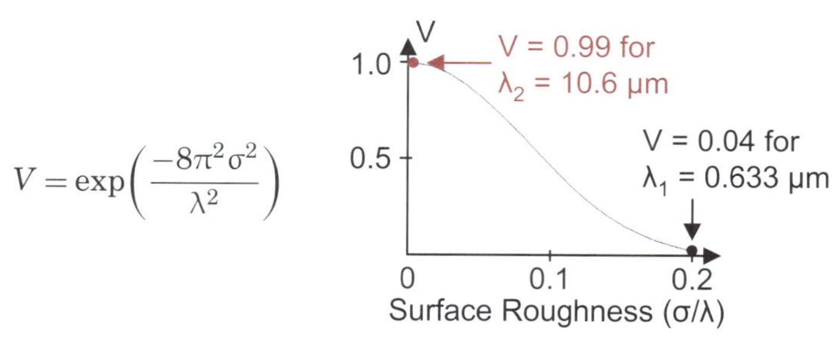

A common source is the CO_2 laser ($\lambda = 10.6$ μm). Due to the infrared wavelength, germanium or zinc selenide optics and a bolometer must be used. This technique can be used in any interferometer configuration, such as a Twyman-Green or Fizeau.

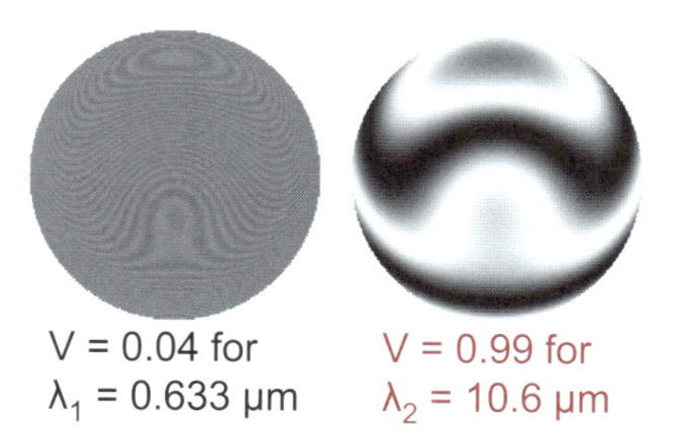

V = 0.04 for $\lambda_1 = 0.633$ μm V = 0.99 for $\lambda_2 = 10.6$ μm

In addition to better contrast, the long-wavelength interferogram has much lower frequency fringes. The ratio of the fringe frequencies is simply the ratio of the two wavelengths.

> Other rough surfaces or surfaces with large departure, such as aspheres, can be tested using this technique.

Smartt Point Diffraction Interferometer

The **point diffraction interferometer** (PDI) is a simple, two-beam interferometer that gives interferograms similar to a Twyman-Green. The wavefront is sent through a pinhole in an attenuating mask. A portion of the incoming beam focuses on the pinhole and is diffracted, creating the reference beam. The test beam passes through the mask and is uniformly attenuated. The attenuation helps to match the irradiance of the reference beam. Tilt in the reference beam is introduced by laterally translating the pinhole while defocus is achieved through longitudinal translation.

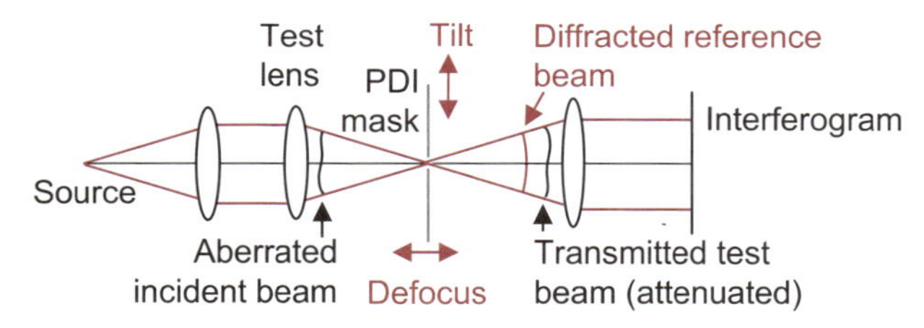

The PDI requires no high-quality optics since every element is common to both beams. The major disadvantage is that the amount of light in the reference beam depends on the position of the pinhole. As the amount of aberration and/or the tilt increases, the fringe visibility drops. Typically, any tilt greater than 5 to 7 fringes makes the visibility unacceptable.

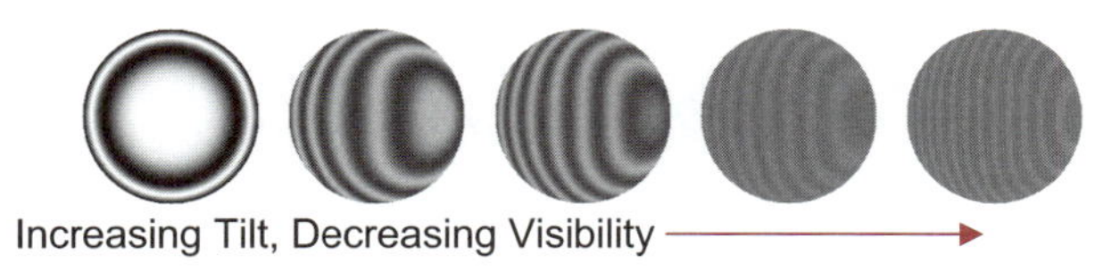

The addition of a pellicle beamsplitter allows the testing of both concave mirrors and lenses tested in double pass.

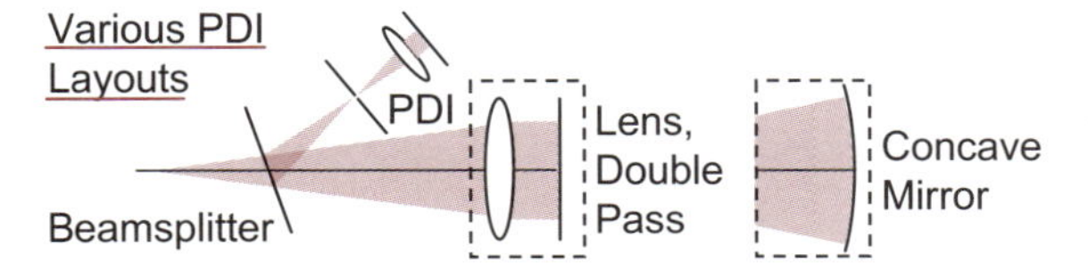

Phase Shifting a PDI

This common path interferometer can be phase shifted with a few modifications. First, a diffraction grating creates the reference (on-axis) and test (first diffraction order) beams, both of which go through the test optic. Second, the attenuating pinhole PDI mask is replaced with an opaque mask with two openings in it. The first is large enough to allow the aberrated off-axis test beam through unaffected. On-axis is a pinhole that cleans up the reference beam as a spatial filter would.

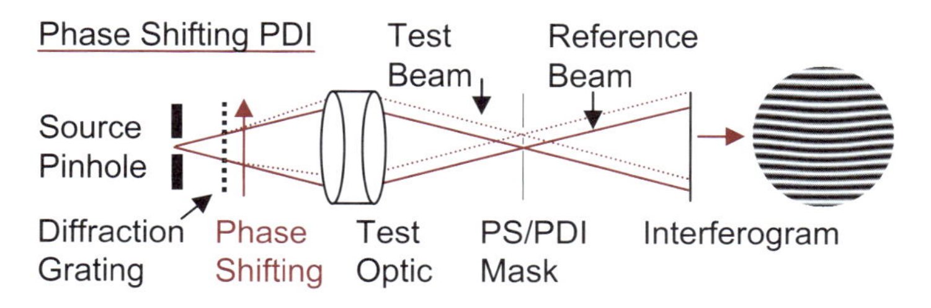

The beam separation in the plane of the phase-shifting PDI mask is determined by the diffraction grating period. The two beams propagate to the interference plane, where the tilt is due to their angular separation. The phase shifting can be done in one of two ways. The diffraction grating can be shifted 1/4 of a grating period Λ laterally to produce a $\pi/2$ phase shift.

$$\phi_{Shift} = \frac{2\pi}{\Lambda}\Delta x$$

Alternatively, the grating period can be designed to introduce enough tilt in the interferogram that a spatial carrier technique can be used on a single image. This is a remarkably simple common-path simultaneous phase shifting interferometer; insensitive to vibrations and requiring no high-quality optics. Having the reference beam on-axis ensures that a wavelength change will not necessitate a shift of the PS/PDI mask.

Sommargren Diffraction Interferometer

This interferometer has the advantage of not using a reference surface and has minimal high-quality auxiliary optics. A low-coherence source is split into two beams, where one beam is delayed a distance L depending on the radius of curvature of the test part. The beams are recombined and transferred via a single-mode fiber to the test part. The delayed beam becomes the reference beam and goes through the imaging lens to the camera. The other beam reflects off the test optic and then off the semi-transparent metallic film coating on the end of the fiber to the camera. Both beams were spatially filtered by transmitting through the single-mode fiber.

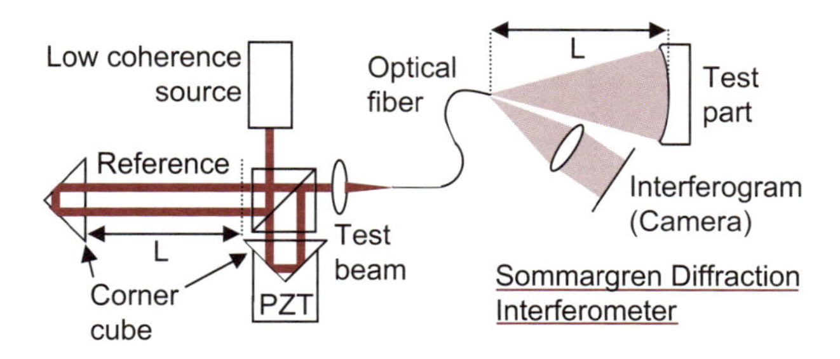

Since the source has low temporal coherence, the portion of the test beam that goes straight to the imaging lens and the portion of the delayed reference beam that reflects off the test part will not interfere with the test and reference beams and will only add a constant irradiance signal. Phase shifting is achieved using a PZT on the corner cube for the test beam.

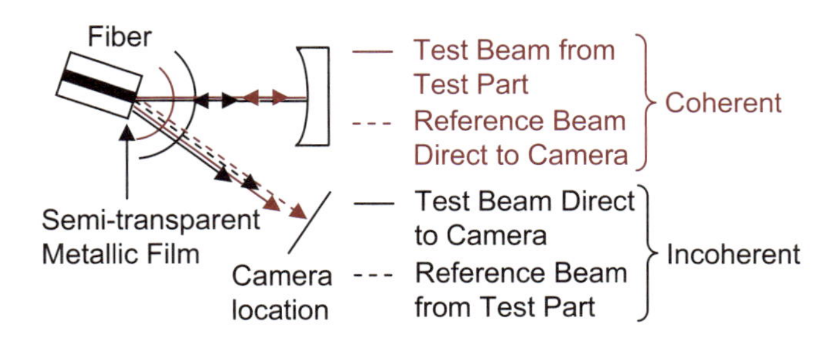

Curved Surfaces, VSWLI

Curved surfaces can also be measured with a **vertical-scanning white light interferometer** (VSWLI), although the test part size is limited to the field of view of the interferometric objective. The test surface slope must be small enough that the reflected light will make it back through the system. This interferometer is not suited for testing lenses in transmission.

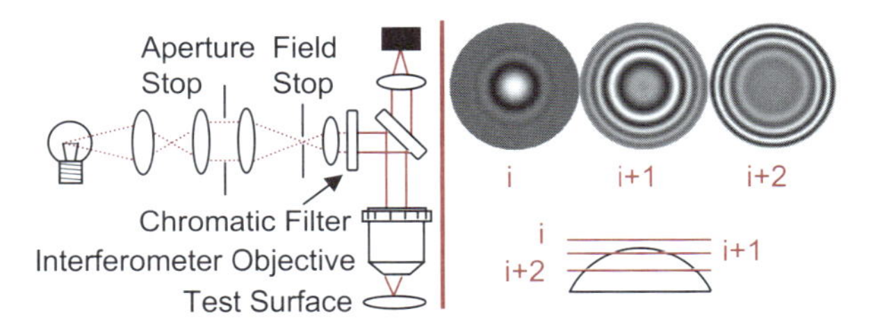

Cylindrical Optics

Testing cylindrical optics interferometrically requires a cylindrical wavefront that somewhat matches the test part. The first method for obtaining such a wavefront is to use a cylinder null lens in a laser-based Fizeau interferometer. This type of null is difficult to make and to characterize.

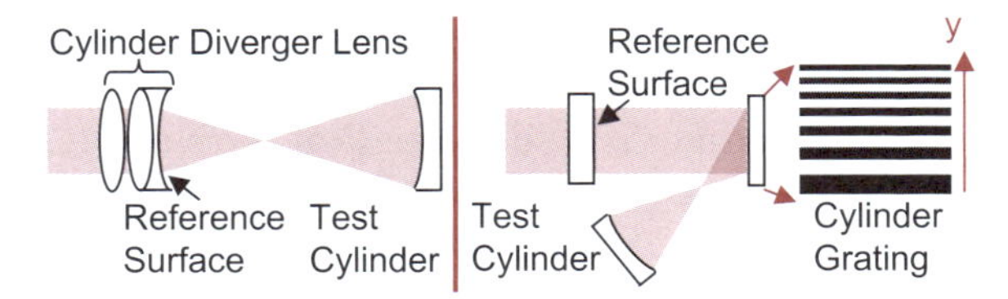

The second approach is to use a cylindrical diffraction grating, which can be thought of as a one-dimensional Fresnel zone plate where the spatial frequency of the grating changes only in one direction.

$$r_m = \sqrt{mf\lambda} \rightarrow y_m = \sqrt{mf\lambda}$$

Here, f is focal length and m is the zone number. The clocking angle of the test part is critical for any method for testing cylindrical surfaces. Both systems can be phase shifted by moving the reference surface with a PZT.

Absolute Measurements: Flats

An interferogram always gives the difference in path length between a reference beam and a test beam. Often the reference surface is assumed to be a flat surface and any wavefront deviation observed in the interferogram is assigned to deviations in the test part. This assumption is only as accurate as the previous characterization of the reference flat. By absolutely characterizing a reference flat, the measurement accuracy is improved because the number of unknown parameters in the interferometer is reduced. By making four measurements between three flat surfaces (A, B, C), x and y-profiles for all three are obtained. These measurements should be performed in a phase-shifting interferometer to improve accuracy.

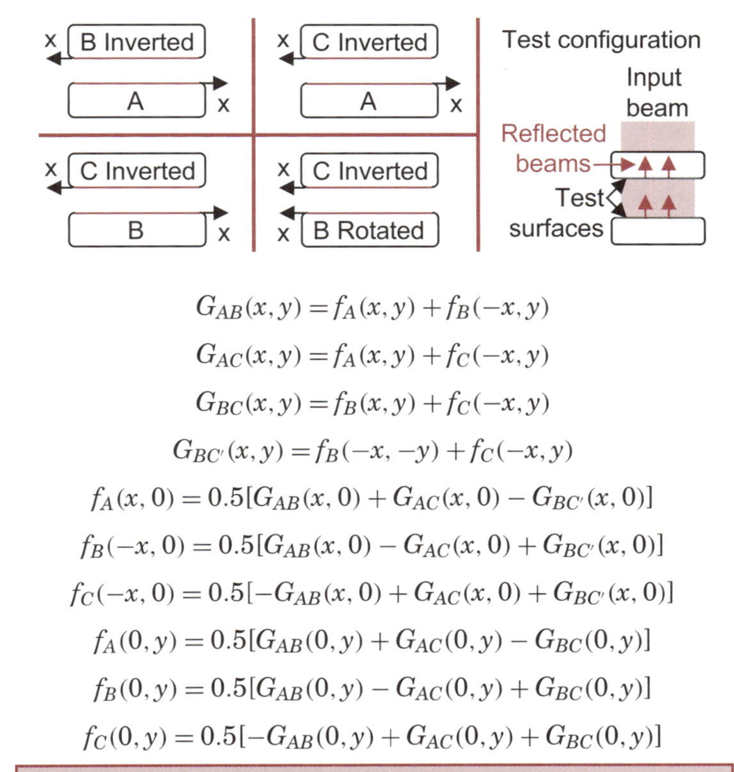

$$G_{AB}(x,y) = f_A(x,y) + f_B(-x,y)$$

$$G_{AC}(x,y) = f_A(x,y) + f_C(-x,y)$$

$$G_{BC}(x,y) = f_B(x,y) + f_C(-x,y)$$

$$G_{BC'}(x,y) = f_B(-x,-y) + f_C(-x,y)$$

$$f_A(x,0) = 0.5[G_{AB}(x,0) + G_{AC}(x,0) - G_{BC'}(x,0)]$$

$$f_B(-x,0) = 0.5[G_{AB}(x,0) - G_{AC}(x,0) + G_{BC'}(x,0)]$$

$$f_C(-x,0) = 0.5[-G_{AB}(x,0) + G_{AC}(x,0) + G_{BC'}(x,0)]$$

$$f_A(0,y) = 0.5[G_{AB}(0,y) + G_{AC}(0,y) - G_{BC}(0,y)]$$

$$f_B(0,y) = 0.5[G_{AB}(0,y) - G_{AC}(0,y) + G_{BC}(0,y)]$$

$$f_C(0,y) = 0.5[-G_{AB}(0,y) + G_{AC}(0,y) + G_{BC}(0,y)]$$

It is not possible to obtain absolute full surface maps of flats.

Absolute Measurements: Spheres

A single spherical surface can be absolutely characterized with three measurements. The addition of the cat's eye position characterizes the interferometer without the test sphere to be measured. The three contributing aberrated wavefronts are due to the test surface, the reference arm, and the diverger lens.

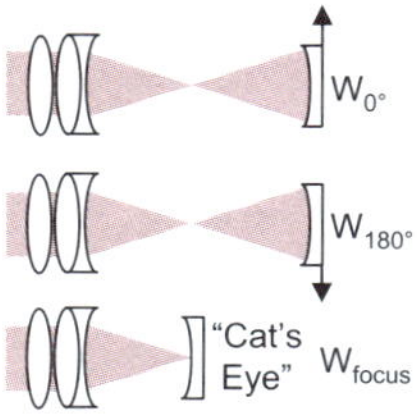

$$W_{0^\circ} = W_{surf} + W_{ref} + W_{div}$$

$$W_{180^\circ} = \overline{W}_{surf} + W_{ref} + W_{div}$$

$$W_{focus} = W_{ref} + \frac{1}{2}(W_{div} + \overline{W}_{div})$$

$$\overline{W} = W \text{ rotated } 180^\circ \text{ about optical axis}$$

$$W_{surf} = \frac{1}{2}(W_{0^\circ} + \overline{W}_{180^\circ} - W_{focus} - \overline{W}_{focus})$$

Absolute Measurements: Surface Roughness

An absolute surface roughness test assumes:

- Surface height is random
- Statistics do not vary over the surface
- Test and reference surfaces are uncorrelated

Any surface roughness on the reference surface will cause errors in the measurement of σ of the test surface.

$$\sigma_{meas} = \sqrt{\sigma_{test}^2 + \sigma_{ref}^2} \quad \sigma_{test} = \sqrt{\sigma_{meas}^2 - \sigma_{ref}^2}$$

This error can be removed by taking many measurement frames N, each time moving the test surface further than the surface correlation length. The test surface roughness effect is reduced by $1/\sqrt{N}$ until the reference surface roughness dominates; the result is a surface map. Subtracting the measured σ_{ref} from any frame gives σ_{test}.

$$\sigma_{test,N} = \frac{1}{\sqrt{N}}\sigma_{test}; \quad \sigma_{meas,N} \cong \sigma_{ref,N}; \quad \sigma_{test} = \sqrt{\sigma_{meas}^2 - \sigma_{ref,N}^2}$$

An easier alternative is the absolute surface test. The test part is moved further than the surface correlation length between two measurements. Only σ_{test} is determined:

$$diff = meas_1 - meas_2; \quad \sigma_{test} = \frac{1}{\sqrt{2}}\sigma_{diff}$$

Aspheric Surfaces

Aspheric surfaces are important to optical systems because they can provide improved performance, fewer optical components, reduced weight, and lower cost. These improvements are due to the greater flexibility and aberration compensation possible with nonspherical surfaces. An asphere is technically any surface that is not spherical, but there are several common aspheric surfaces found in optical systems. Conics are one type of rotationally symmetric aspheric surface defined by a **conic constant** κ, the **sag** of which is given by

$$s(r) = \frac{Cr^2}{1+\sqrt{1-(1+\kappa)C^2r^2}}; \quad C = \frac{1}{R}; \quad r^2 = x^2 + y^2$$

C is the surface curvature, R is a base radius of curvature, and r is the radial coordinate. Familiar surfaces with two foci (F_1 and F_2) are obtained for certain values of κ.

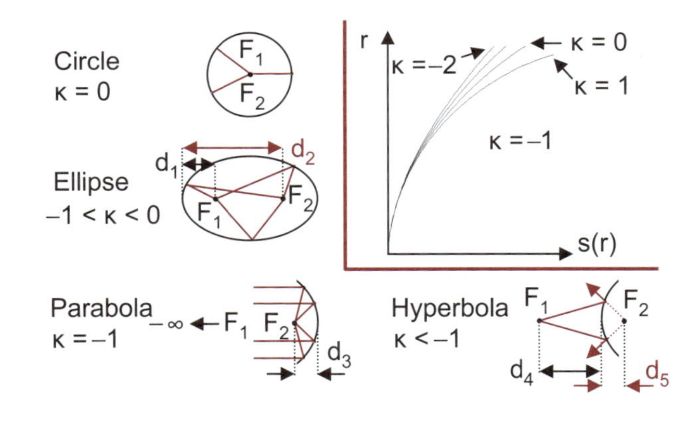

$$d_{1,2} = \frac{R}{\kappa+1}(1 \pm \sqrt{-\kappa}); \quad d_3 = \frac{R}{2}; \quad d_{4,5} = \frac{R}{\kappa+1}(\sqrt{-\kappa} \pm 1)$$

The general asphere contains the conic term and higher-order rotationally symmetric terms. To the level of third-order aberrations, all aspherics with power are indistinguishable from conics; higher-order terms only affect higher-order aberrations.

$$s(r) = \frac{Cr^2}{1+\sqrt{1-(1+\kappa)C^2r^2}} + A_4r^4 + A_6r^6 + A_8r^8 + \cdots$$

Aspheric Testing

The difficulty of testing an asphere is determined by the slope of the aspheric departure from the best fit sphere (the derivative of the OPD). Steeper slopes can lead to high-frequency fringes, retrace errors, and even **vignetting**. These issues are avoided for null testing.

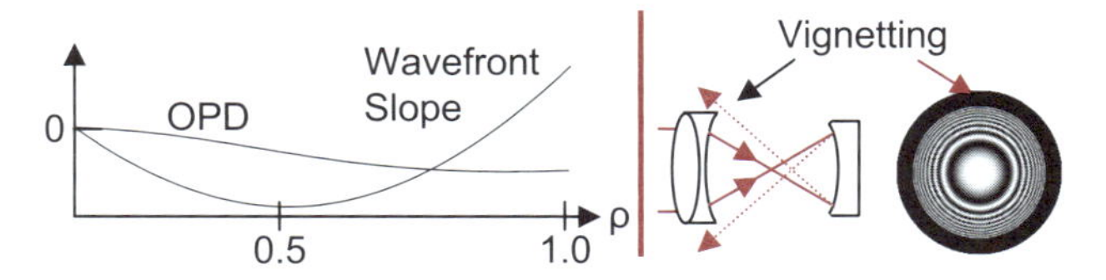

There are two categories of aspheric testing methods; null tests and non-null tests. Null tests

- usually require part specific optical components
- are well suited for testing many of the same part

Conventional Null Optics

Testing aspheres with conventional null optics involves a part-specific lens system to create a wavefront that matches the desired surface of the asphere under test. The lens system usually will consist of spherical surfaces that can be characterized in a standard interferometer. The null lens can be combined with a standard phase-shifting Twyman-Green or Fizeau interferometer.

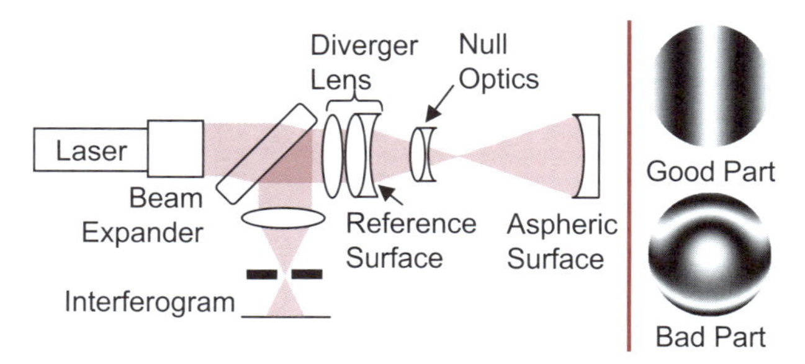

If the test part is correct, the interferogram will contain only defocus and tilt. As with testing other surfaces, part errors are evident in the interferogram. Having a new lens or lenses made for each aspheric test surface is time consuming and expensive, but the system can be well characterized and understood.

Hyperboloid Null Tests

Specific convex hyperboloid surfaces that are transparent have a plane second surface and an index of refraction $n = \sqrt{-\kappa}$ can be tested in the **Cartesian** configuration. **Meinel** suggested testing the surface through the glass, from the back. The spherical back surface is slightly convex, giving equal conjugates. For a plane back surface, a double-pass test can be performed by placing a small spherical mirror near one of the focus points.

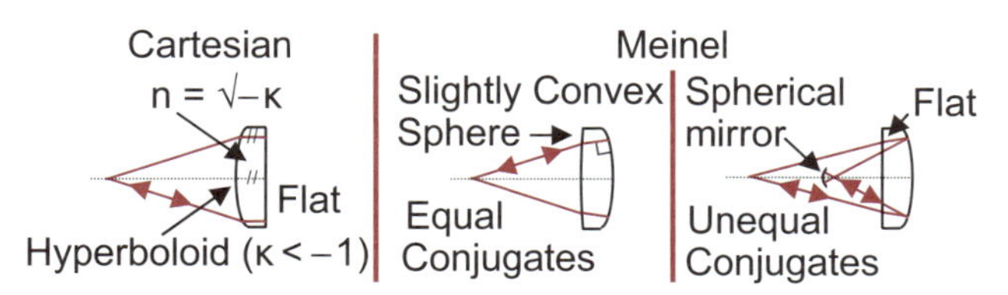

Hindle tests are more common, where a concave spherical mirror is used to test a convex hyperboloid. The biggest problem is that the **Hindle sphere** can become very large. If m is the magnification of the hyperboloid's conjugates and B is the permissible obscuration ratio, the diameter of the Hindle sphere is D_{HS}.

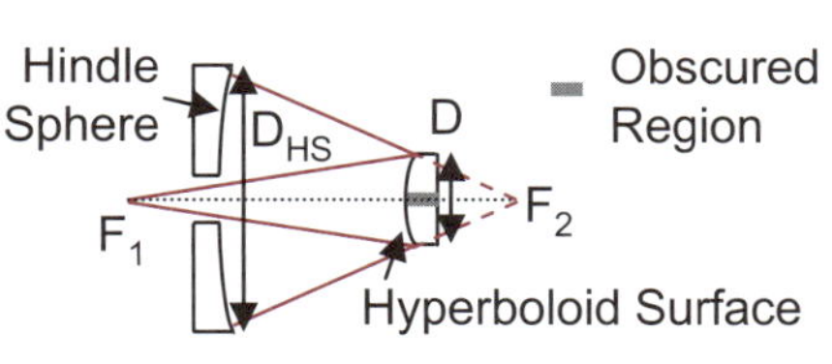

$$D_{HS} = \frac{D(m+1)}{mB+1}$$

The **Simpson-Oland-Meckel test** uses a partially reflective Hindle sphere that also transmits the test beam. Since there is no central obscuration, the sphere can be much smaller and the entire surface is tested. Any OPD variations in transmission through the Hindle surface are measured with a calibration sphere.

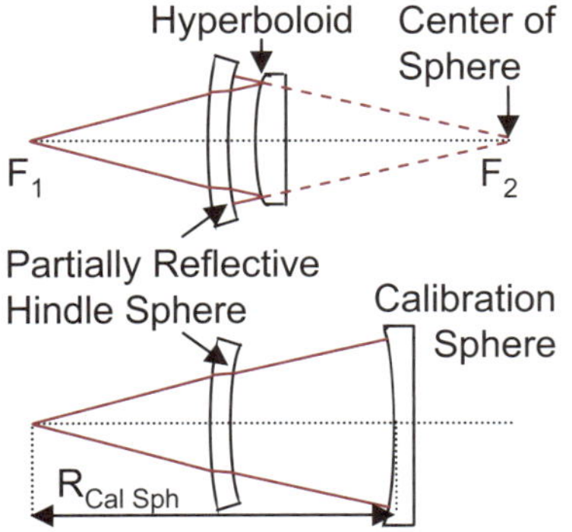

Offner Null

A refractive **Offner null** compensator can be used for concave mirrors, where lens A images a point source at the center of curvature of the test surface. The field lens F helps combat fifth-order spherical aberration. For high-accuracy tests, slight index variations in lens A can lead to large OPD variations in the interferogram. The reflective Offner null compensator uses a spherical mirror at a magnification other than −1 to introduce spherical aberration opposite to that of the test part. These setups can be used for conics with $\kappa < 0$.

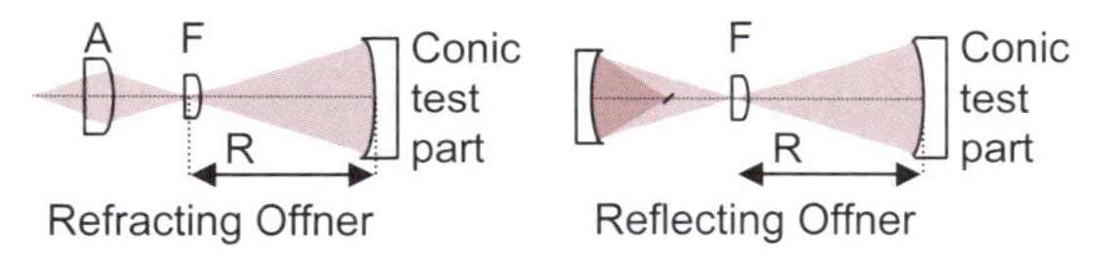

Parabaloid Surface Tests

Concave paraboloids are tested in double pass with a flat mirror as large as the test part. The configuration varies slightly if the test surface has a central obscuration.

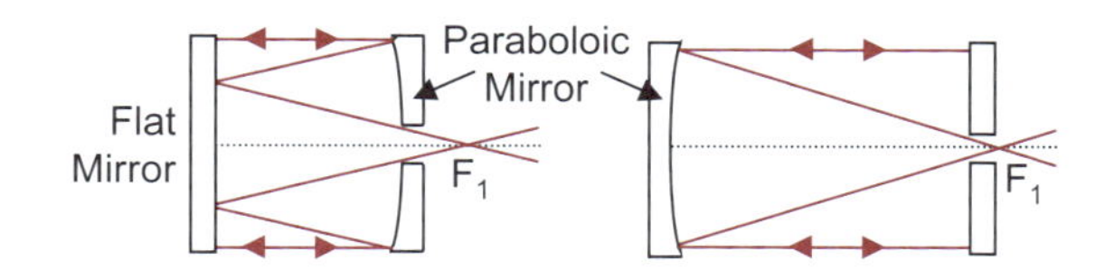

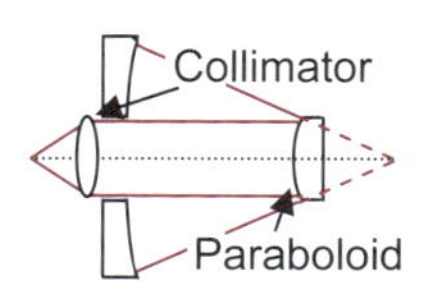

Convex parabaloids can be tested using the Hindle test by placing a collimating lens near the plane of the Hindle sphere. The obscuration diameter is that of the test part.

Elliptical Surface Tests

An elliptical surface is tested with a spherical wave originating at one focus and emerging at the other. In an interferometer, a small spherical mirror can be placed near the innermost focus to return the light in a double-pass test configuration.

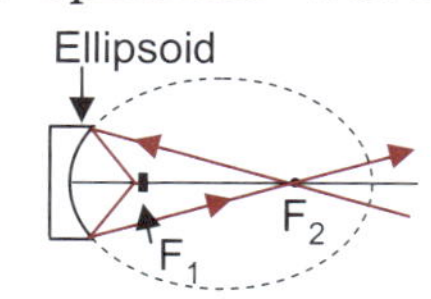

Holographic Null Optics

Computer-generated holograms (CGHs) are used in interferometric optical testing to measure aspheres. The most common configuration is to use a CGH in place of a conventional refractive null lens. It is often easier to design and fabricate a CGH to test an aspheric part than it is to design and build a refractive null. The advantage of this configuration is that the CGH can be added to an existing Fizeau or Twyman-Green interferometer. The disadvantage is that any phase variations in the CGH substrate are unique to the test wavefront. Therefore, the substrate must be highly accurate or well characterized.

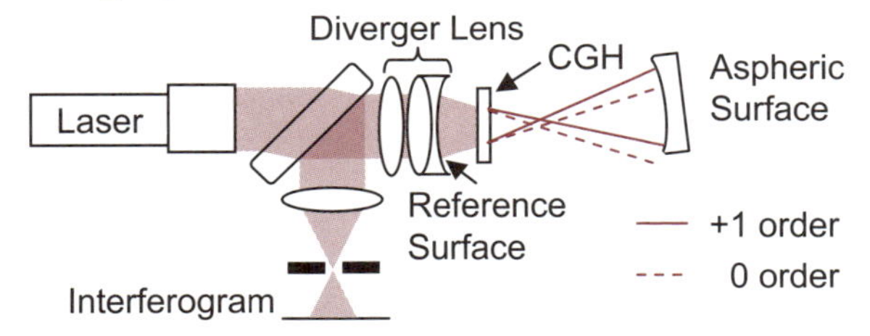

Position the test part to normally reflect the +1 order.

Common Path CGH Configuration

In this setup, both the test and reference beams travel through the CGH, eliminating the sensitivity to substrate errors. Usually the +1 order is used to correct the test beam while the unaltered 0 order of the reference beam is allowed through by the spatial filter. The major disadvantage to this setup is that it requires a custom interferometer; it is difficult to place a CGH at this location in a commercial Fizeau interferometer. Reverse raytracing of the test beam may also be required.

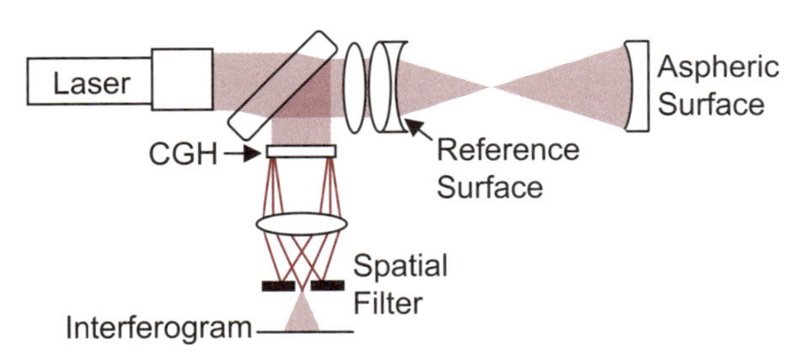

CGH Basics

A CGH looks like a binary representation of an interferogram. Each line in the CGH adds $m\lambda$ of OPD and changes the wavefront slope by $\sin(\theta) = m\lambda/\Lambda$, where Λ is the local line spacing of the CGH. The 0 order is unaltered by the CGH. Below, the +1 order is allowed to transmit.

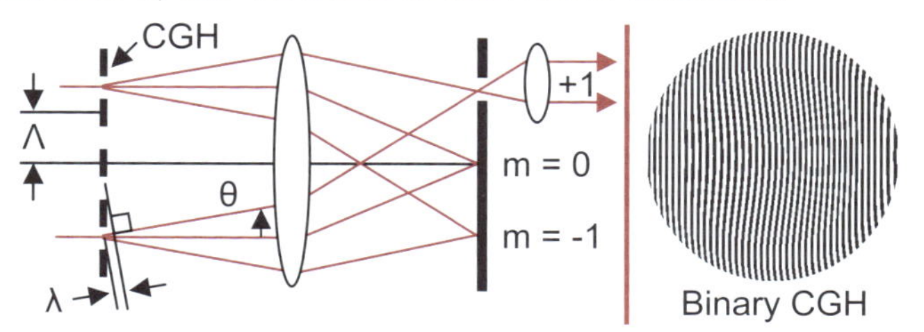

Amplitude vs. Phase CGH

CGHs for optical testing are usually binary. The transmission of an amplitude grating alternates between 0 and 1, resulting in about 10% diffraction efficiency (η) in the ± 1 orders and 25% in the 0 order. The two phase levels of a phase CGH differ by $\lambda/2$ of OPD, or 180° of phase. The efficiency is 40% for the ± 1 orders and nominally 0% for the 0th order.

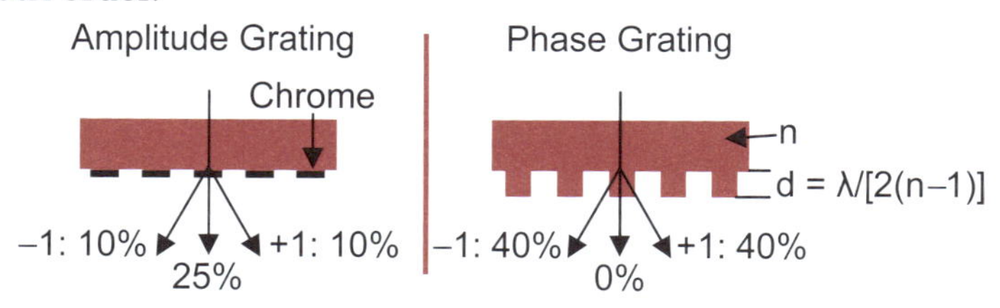

For diffraction order $m > 0$, the optimal diffraction efficiency is:

$$\eta_{Amplitude} = 0.25 \cdot \text{sinc}^2(m/2)$$

$$\eta_{Phase} = \text{sinc}^2(m/2)$$

Phase CGHs are more common. The double-pass nature of placing the CGH in the test beam requires using a phase CGH to get enough light.

CGH fabrication uses lithographic techniques that were developed for the semiconductor industry.

CGH Design Guidelines

The goal of a CGH is to create a well-known aspheric wavefront from a spherical wavefront. After choosing a test configuration, the accuracy, size, and cost must be balanced using the following rules of thumb:

Line spacing:
- 10 μm is routine. 0.1 μm error, 0.01λ wavefront error
- 1 μm is harder. More errors, poorer accuracy

Size:
- 50 mm CGH is low risk, low cost
- Up to 150 mm possible; expensive to fabricate

The CGH needs to be large enough to fit the details without pushing the line spacing limits. In order to separate the multiple diffraction orders, a tilt carrier must be incorporated into the CGH. There should be enough tilt to separate the orders, but too much tilt means smaller line spacing and a higher sensitivity to errors.

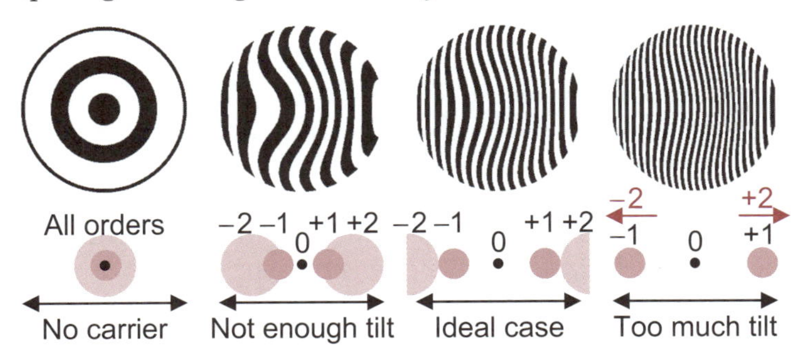

The phase imparted by the CGH can be modeled as a phase function in lens design code.

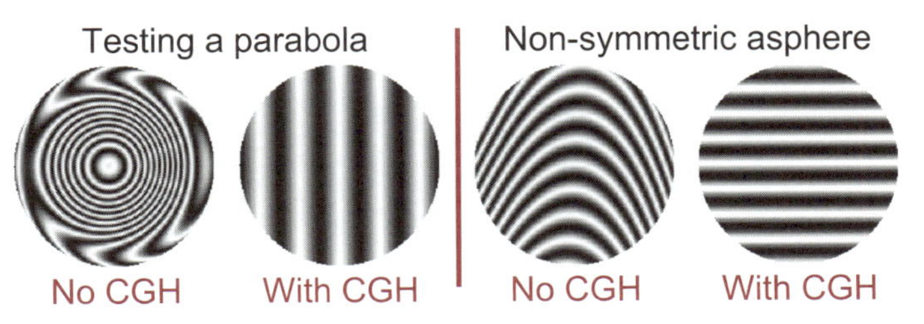

CGHs are now commercially available. With care, they can be quite accurate. They are flexible enough to help test things that may not be testable otherwise.

Non-Null Tests

The goal of a non-null interferometer for testing aspheres is to not have to rely on a part-specific null lens or CGH. There are 3 general requirements for any interferometric test; many non-null configurations push these limits:

- must get light back into the interferometer
- must be able to resolve the fringes
- must know precisely the optical test setup

A null interferometer strives to create a wavefront that matches the test surface and is normally incident everywhere. The reflected light retraces the incident path, and a null interferogram results.

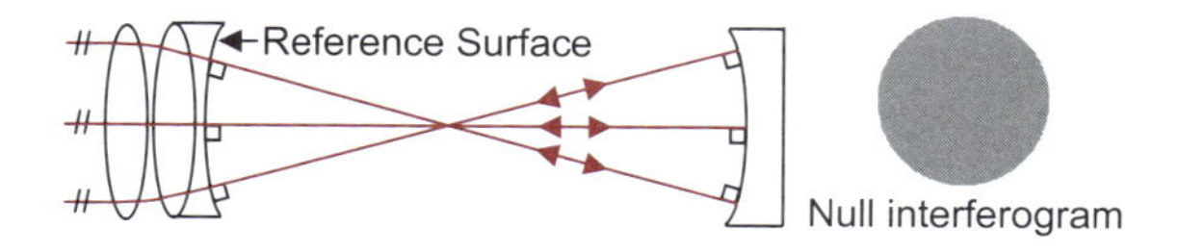

If the test surface and incident wavefront do not match, interference fringes result. The light is no longer normally incident, and does not follow the same path back through the interferometer. The wavefront is altered on the return path, adding part-dependent **induced aberrations** (**retrace errors**) to the test wavefront. The amount of induced aberration increases with increasing part deviation. Any deviation from null will add some induced aberration.

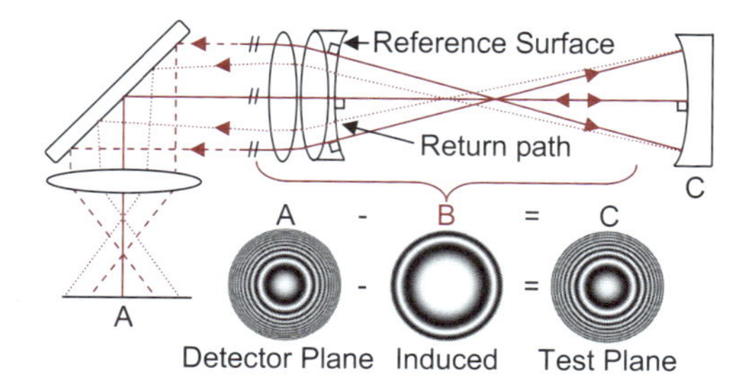

In order to obtain the true test surface, the induced aberrations must be removed. This can be done by **reverse ray-tracing**. The wavefront at the detector is found from the interferogram. The rays are traced back through the system to the test plane to determine the surface figure.

Reverse Raytracing

Reverse raytracing requires precise knowledge of the interferometer setup. Some parameters can be accurately measured with independent tests, such as the surface figure for the diverger lens and the refractive index of the lenses and beamsplitters. Absolute part locations (x, y, z) can be more difficult. **Reverse optimization** (RO) is a method for finding the overall system description. Nominal system information is used as a starting point, and a series of known test parts are measured. The system parameters are then variables in the optimization to find the best description of the interferometer.

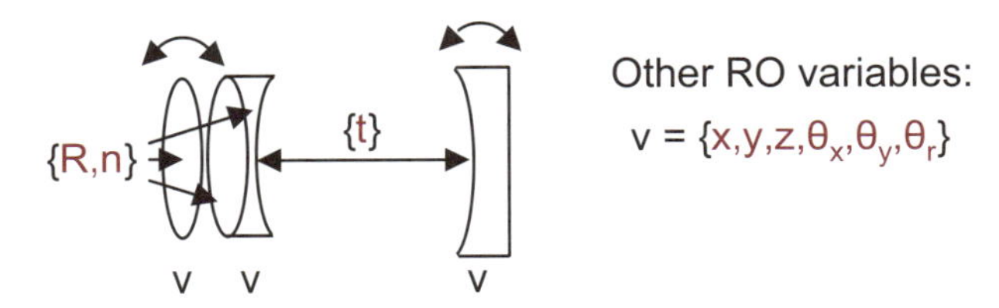

High-Density Detector Arrays

Testing aspheres with a non-null interferometer typically leads to interferograms with high-frequency fringes. A high-density detector array is capable of digitizing these rapidly varying interferograms. As long as there are at least two pixels per fringe, the **Nyquist limit** (ξ_{Ny}) is satisfied and the interferogram is properly captured. Accurate tests require reverse raytracing to back out the induced aberrations.

$$\xi_{Ny} = \frac{1}{2x_s}$$

x_p
Nyquist limit
Pixels
$\varphi \approx$ 0° 180° 0° 180°

Sub-Nyquist Interferometry

Sub-Nyquist interferometry is a non-null interferometric technique requiring fewer than two pixels per fringe. Most CCD detector arrays have a fill factor (**G-factor**) close to 1; that is, the pixel width (x_w) is almost as large as the pixel spacing (x_s).

$$\xi_{Ny} = \frac{1}{2x_s} \quad \xi_c = \frac{1}{x_s} = 2\xi_{Ny}$$

$$G = \frac{x_w}{x_s} \quad \xi_{c,sa} = \frac{1}{Gx_s}$$

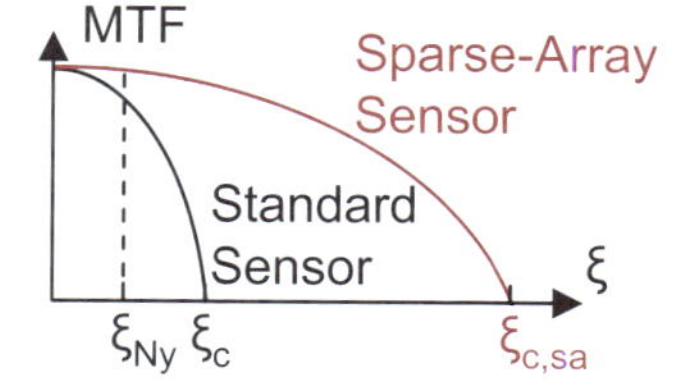

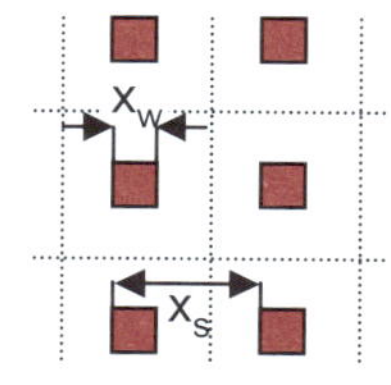

A detector array is modified by shrinking the pixel width relative to the pixel spacing, resulting in a **sparse array** detector. This allows fringes past the Nyquist limit to alias. PSI is used to find the modulo 2π phase at each pixel.

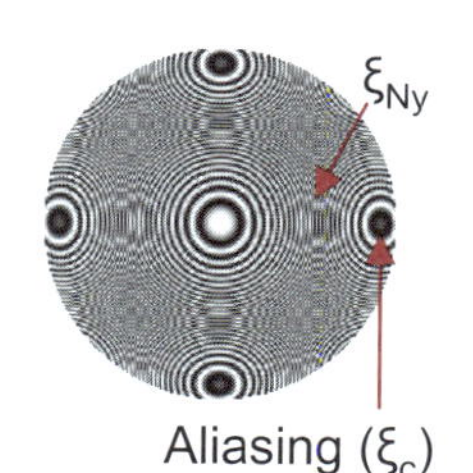

Recall that PSI is usually unwrapped by adding or subtracting 2π to minimize the phase difference between adjacent pixels. Sub-Nyquist interferometry allows phase changes greater than π between adjacent pixels, so unwrapping the phase must take into account the local derivative of the wavefront. The assumption is made that the first and second derivatives are continuous, which is valid for most optical surfaces, including aspheres. The unwrapping must begin in a region with no aliasing.

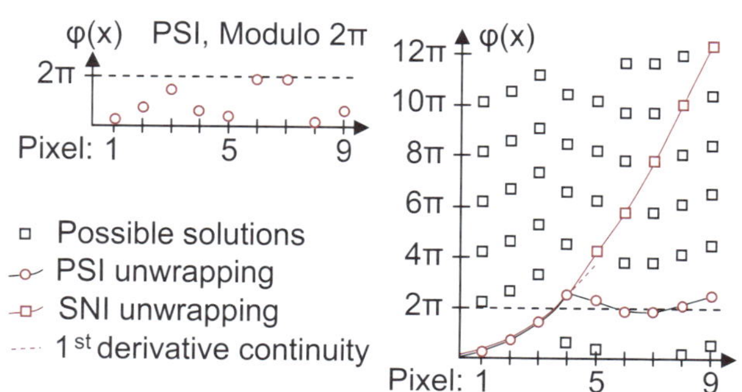

Long-Wavelength Interferometry

This technique changes the wavelength of the light source for the interferometer. A common source is a carbon dioxide (CO_2) laser with $\lambda = 10.6$ µm, which has good coherence properties. The resulting interferogram has many fewer fringes than would be obtained using a visible laser, so the Nyquist limit is no longer in danger of being violated for aspheres. The major drawback is reduced sensitivity. Germanium or zinc selenide optics and a bolometer must be used.

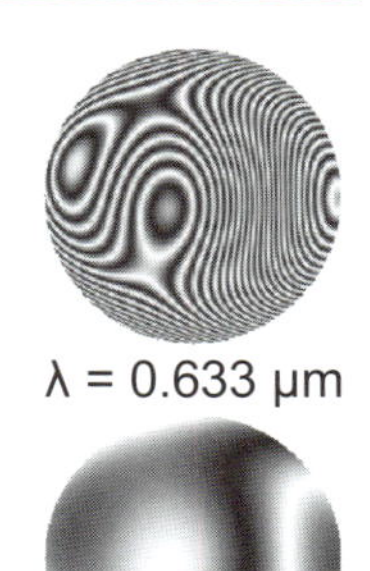

λ = 0.633 µm

λ = 10.6 µm

Two-wavelength holography (TWH) is another reduced sensitivity test, but one that does not require exotic glasses and detectors. A hologram of the test part interfered with the reference beam is recorded at λ_1. The same test is performed at λ_2 with the hologram in place. The resulting interferogram is the same as what would be obtained using the equivalent wavelength, λ_{eq}.

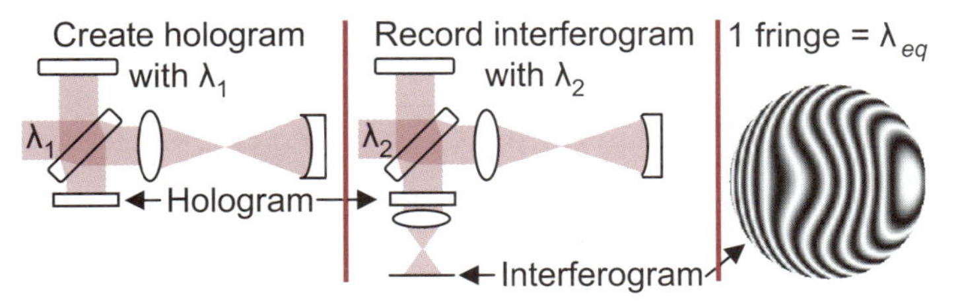

Similarly, **two-wavelength interferometry** can be used to test aspheres. Two interferograms are recorded at λ_1 and λ_2. A computer calculates the difference between the two measurements. The wavefront is sufficiently sampled if there would be at least two pixels per fringe for a wavelength of λ_{eq}. Chromatic aberration is an issue for both of these techniques.

$$\lambda_{eq} = \frac{\lambda_1 \lambda_2}{|\lambda_1 - \lambda_2|}$$

λ_1 = 458 nm λ_2 = 488 nm λ_{eq} = 7.4 µm

Non-Interferometric Testing

The following techniques are common non-interferometric optical testing methods that all measure the slope of the wavefront error rather than the wavefront error itself, which is what most interferometers measure.

The classic **Hartmann test** uses a plate with an array of holes near an illuminated test surface. Photographic plates or detector arrays are placed in the converging beam. If one detector is used, the positions of the holes in the screen and the separation of the screen and detector plane must be known. If two plates are used, only the distance between the two detector planes must be known.

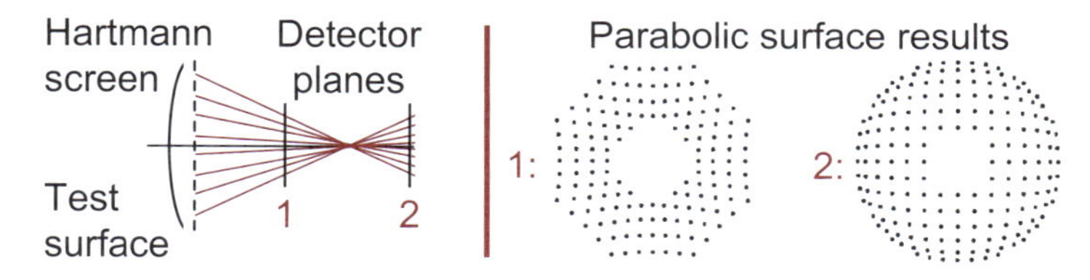

The more common test used today is the **Shack-Hartmann test**, where the holes in the screen are filled in with small lenses, creating a **lenslet array**. Usually, the transmitted or reflected test wavefront is incident on the lenslet array. A detector is placed in the back focal plane of the lenslets. A plane wave will produce a uniform array of spots, while any slope deviations in the wavefront will displace the spots. The spot locations are used to find the wavefront slope in the plane of the lenslets. This dynamic range and accuracy of this test varies with the focal length of the individual lenslets and the number of lenslets in the array. This test is used for optical testing and in **adaptive optics** for active atmospheric correction in telescopes.

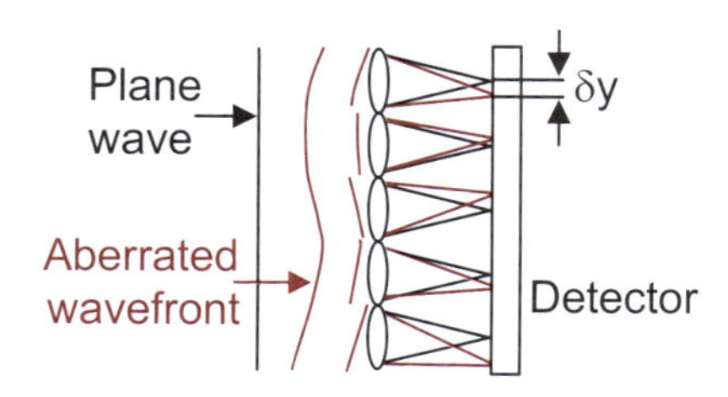

Foucault (Knife-Edge) Test

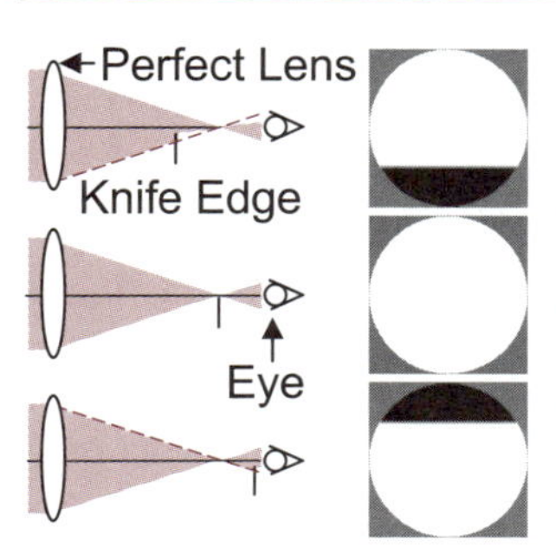

This test is one of the oldest and most common tests for determining longitudinal and transverse aberrations. A **knife edge** is placed near the focus and passed through the image of a point or slit source. The shadow, observed by the eye (shown) or on a screen, gives information about the aberration content. A perfect lens will have one image point that darkens almost instantaneously when the knife edge passes though the image. These shadow patterns are based off of geometrical analysis; diffraction will blur out the edges.

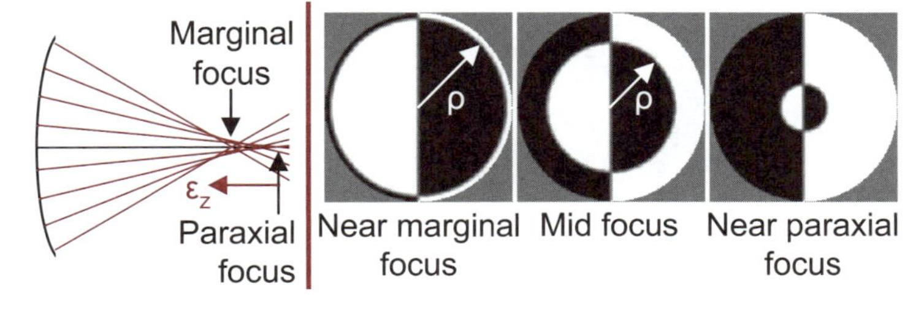

Spherical aberration can be determined using this test. The boundary of the geometrical shadow with normalized pupil coordinates and the knife edge on the optical axis is a vertical line ($y = 0$) and a circle of radius ρ, where ε_z is the axial distance from **paraxial focus**, R is the radius of curvature of the wavefront, and r_p is the pupil radius. By measuring ρ as a function of ε_z, W_{040} can be determined.

$$\rho = \sqrt{x^2 + y^2} = \sqrt{\frac{-\varepsilon_z r_p^2}{4R^2 W_{040}}}$$

The advantage of the knife-edge test is its simplicity. No accessory optics are required, the whole surface is measured at once, and it is a sensitive test. The disadvantage is that it is sensitive to slopes rather than heights and only measures in a single direction with a single orientation of the knife edge.

Foucault (Knife-Edge) Test (cont'd)

The knife edge is most often used to measure the zonal focus of different parts of an optical surface so the optician can determine the high and low surfaces parts.

An improved alternative to the knife edge is to use a phase delay knife edge, where both sides transmit with a phase difference between the two halves of 180°. The diffraction pattern is then symmetric, and the boundary centers are easier to determine.

The **wire test** is the same as the Foucault test except the knife edge is replaced with a wire, or inversely, a slit. The wire can simply be a strand of hair. The wire test is inferior for qualitative data but superior for obtaining quantitative data since the wire produces a symmetric shadow. It is easier to determine the center of the shadow.

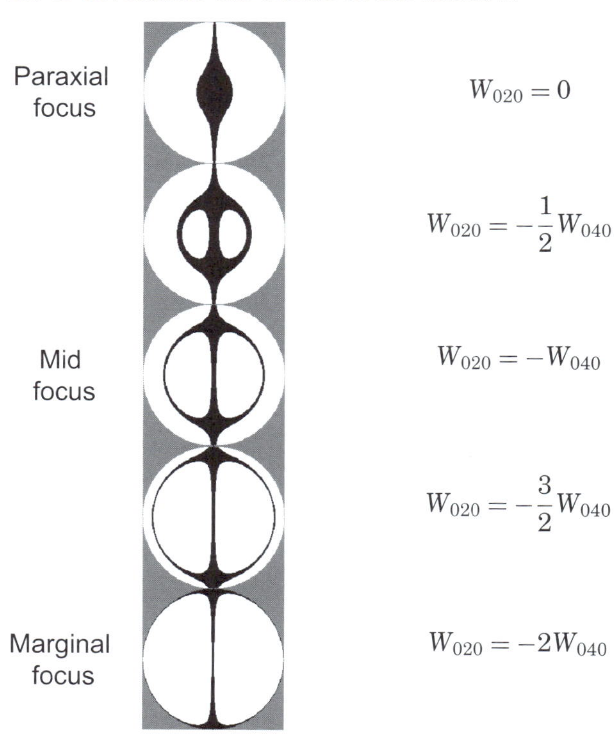

$$W_{020} = 0$$

$$W_{020} = -\frac{1}{2}W_{040}$$

$$W_{020} = -W_{040}$$

$$W_{020} = -\frac{3}{2}W_{040}$$

$$W_{020} = -2W_{040}$$

Ronchi Test

A **Ronchi test** uses a low-frequency grating, called a **Ronchi ruling**, in place of the knife edge in a Foucault test or the wire in the wire test. For a perfect lens, the pattern observed is straight lines, with fewer lines closer to focus.

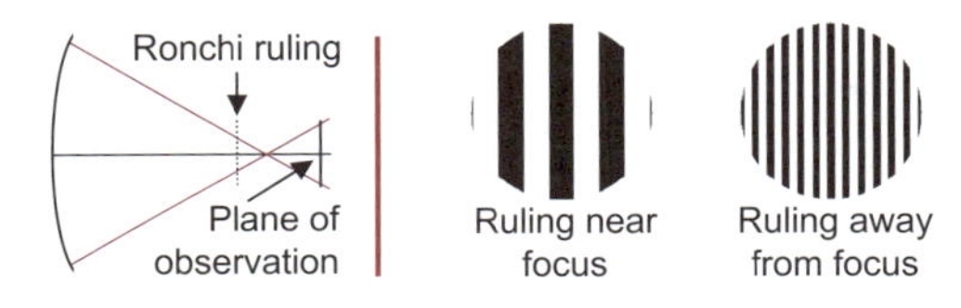

For a wavefront with third-order spherical aberration (W_{040}), the Ronchi patterns are as follows. These examples neglect diffraction patterns.

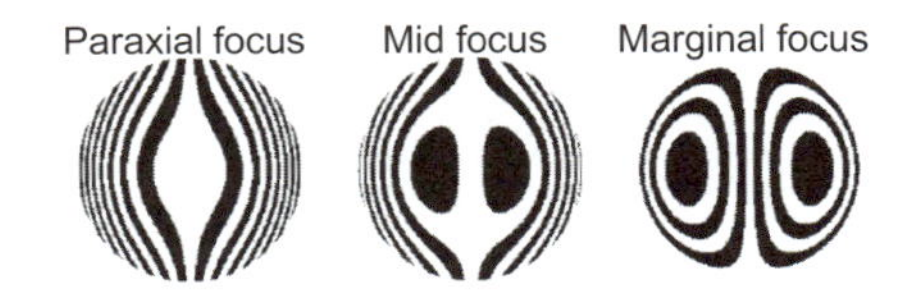

The Ronchi ruling is usually illuminated by a diffuse source from behind and acts as a multiple slit source. A white light source may be used. The Ronchi ruling acts as a diffraction grating, creating multiple images of the test surface. The orders from coarse gratings (fewer than 10 lines/mm) will overlap and cause only a slight perturbation of the shadow pattern. Orders from high-frequency gratings (more than 100 lines/mm) will likely be separated. The patterns obtained from using middle-frequency gratings are difficult to analyze.

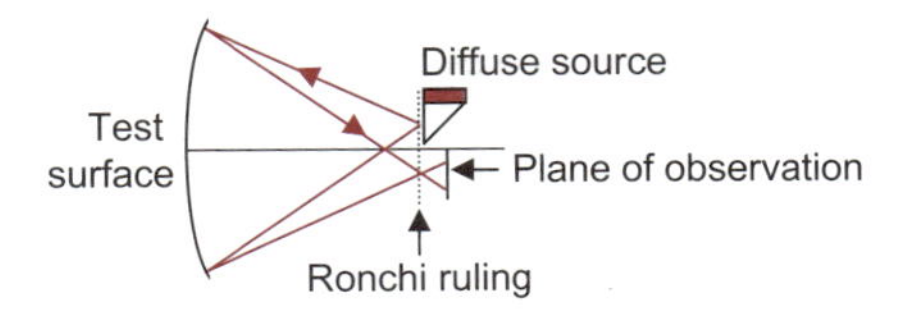

The advantages of the Ronchi test are its simplicity and the fact that white light can be used. Unfortunately, the diffraction effects are very troublesome, limiting the accuracy of this test.

Equation Summary

Two-beam interference equation:

$$I(x,y) = I_1 + I_2 + 2\sqrt{I_1 I_2}\cos(\phi_1 - \phi_2)$$

$$I(x,y) = A_1^2 + A_2^2 + 2A_1 A_2 \cos(\phi_1 - \phi_2)$$

General equations:

$$\upsilon = \frac{c}{\lambda} \quad n(\lambda) = \frac{c}{\upsilon}$$

$$OPL = \int_a^b n(s)ds \quad OPL = nt$$

$$OPD = OPL_1 - OPL_2 \quad (\phi_1 - \phi_2) = \frac{2\pi}{\lambda} OPD$$

Coherence:

$$L_c = \frac{\lambda_c^2}{\Delta\lambda} \quad t_c = \frac{L_c}{c}$$

Visibility:

$$V = \frac{I_{\max} - I_{\min}}{I_{\max} + I_{\min}}$$

Airy disk diameter and numerical aperture:

$$D = \frac{1.22\lambda}{NA} \quad NA = n\sin(\theta)$$

Snell's law:

$$n_1 \sin\theta = n_2 \sin\theta'$$

Beam displacement from a tilted plate:

$$d = t\sin(\theta)\left[1 - \sqrt{\frac{1-\sin^2(\theta)}{n^2 - \sin^2(\theta)}}\right] \approx t\theta\left(\frac{n-1}{n}\right); \quad \theta \text{ in radians}$$

Grating equation:

$$\sin(\theta_d) - \sin(\theta_i) = \frac{m\lambda}{\Lambda}$$

Fresnel reflectance coefficient:

$$\rho = \left(\frac{n_2 - n_1}{n_2 + n_1}\right)^2$$

Equation Summary (cont'd)

Fringe to height error conversions:

$$\text{Twyman-Green, Fizeau: 1 fringe} = \frac{\lambda}{2}\ \text{Surface height}$$

$$\text{Mach-Zehnder: 1 fringe} = \frac{\lambda}{2 \cdot \cos(45)} = \frac{\lambda}{\sqrt{2}}\ \text{Surface height}$$

Height error of a bump or hole:

$$h = \frac{\lambda}{2}\frac{\Delta}{S}$$

Radius of curvature, Newton's rings:

$$R = \frac{r_m^2}{\lambda\left(m + \frac{1}{2}\right)}$$

Lateral shearing interferometer:

$$W(x + \Delta x, y) - W(x, y) = m\lambda = \left[\frac{\delta W(x, y)}{\delta x}_{\text{Avg.over Shear}}\right]\Delta x$$

Shear from a Savart plate:

$$\text{Shear} = \sqrt{2}\frac{n_e^2 - n_o^2}{n_e^2 + n_o^2}t$$

Shear angle from a Wollaston prism:

$$\theta = 2(n_e - n_o)\tan\alpha$$

Radial shear:

$$R_s = \frac{S_1}{S_2}$$

Gaussian imaging equation:

$$\frac{1}{z'} = \frac{1}{z} + \frac{1}{f}$$

Wavefront aberration coefficients:

$$x_P = \rho\sin\theta \quad y_P = \rho\cos\theta \quad \rho = \sqrt{x_P + y_P}$$

$$W = \sum_{L,K,N} W_{IJK} H^I \rho^J \cos^K\theta; \quad I = 2L + K,\ J = 2N + K$$

Equation Summary (cont'd)

Zernike polynomials:

$$x_P = \rho\cos\theta' \quad y_P = \rho\sin\theta' \quad \rho = \sqrt{x_P + y_P}$$

$$g[\theta' + \alpha] = g[\theta']g[\alpha] \quad g[\theta'] = e^{\pm im\theta'}$$

RMS wavefront error:

$$\sigma^2 = \frac{1}{\pi}\int_0^{2\pi}\int_0^1 [\Delta W(\rho,\theta) - \overline{\Delta W}]^2 \rho d\rho d\theta = \overline{\Delta W^2} - (\overline{\Delta W})^2$$

Moiré:

$$C = \frac{\Lambda_1\Lambda_2}{\sqrt{\Lambda_2^2\sin^2 2\alpha + (\Lambda_2\cos 2\alpha - \Lambda_1)^2}}$$

$$\alpha = 0, \quad \text{then: } C = \Lambda_{beat} = \frac{\Lambda_1\Lambda_2}{|\Lambda_2 - \Lambda_1|}$$

Phase-shifting interferometry:

$$I(x,y) = I_{dc}(x,y) + I_{ac}(x,y)\cos[\phi(x,y) + \phi(t)]$$

Continuous phase shifting:

$$\phi(t) = 2\pi\Delta\upsilon t$$

Frequency shifting source:

$$\phi = \upsilon\frac{2\pi}{c}(2\cdot OPD)$$

$$\Delta\phi = \upsilon\frac{2\pi}{c}(2\cdot OPD)$$

Phase shifting:

$$I(x,y) = I_{dc}(x,y) + I_{ac}(x,y)\cos[\phi(x,y) + \phi(t)]$$

$$I_1(x,y) = I_{dc} + I_{ac}\cos[\phi(x,y)] \quad \phi(t) = 0\ (0^\circ)$$

$$I_2(x,y) = I_{dc} - I_{ac}\sin[\phi(x,y)] \quad \phi(t) = \pi/2\ (90^\circ)$$

$$I_3(x,y) = I_{dc} - I_{ac}\cos[\phi(x,y)] \quad \phi(t) = \pi\ (180^\circ)$$

$$I_4(x,y) = I_{dc} + I_{ac}\sin[\phi(x,y)] \quad \phi(t) = 3\pi/2\ (270^\circ)$$

Three step algorithm:

$$\phi = \tan^{-1}\left(\frac{I_3 - I_2}{I_1 - I_2}\right)$$

Equation Summary (cont'd)

Four step algorithm:

$$\phi = \tan^{-1}\left(\frac{I_4 - I_2}{I_1 - I_3}\right)$$

Schwider-Hariharan (5-step) algorithm:

$$\phi = \tan^{-1}\left[\frac{2(I_2 - I_4)}{2I_3 - I_5 - I_1}\right]$$

Average phase shift between frames:

$$a(x,y) = \cos^{-1}\left[\frac{1}{2}\frac{I_5(x,y) - I_1(x,y)}{I_4(x,y) - I_2(x,y)}\right]$$

Fringe contrast (visibility) degradation, phase ramping:

$$I_i = I_0\left\{1 + \Gamma\,\mathrm{sinc}\left(\frac{\Delta}{2}\right)\cos[\phi(x,y) + \phi_i]\right\}; \quad \mathrm{sinc}(x) = \frac{\sin(x)}{x}$$

Spatial synchronous methods:

$$I(x,y) = I_{dc}(x,y) + I_{ac}(x,y)\cos[\phi(x,y) + 2\pi fx]$$

$$\mathrm{Ref}\ 1(x,y) = \cos(2\pi fx) \quad \mathrm{Ref}\ 2(x,y) = \sin(2\pi fx)$$

$$\phi(x,y) = \tan^{-1}[-Sm(I * \mathrm{Ref2})/Sm(I * \mathrm{Ref1})]$$

Quantization error:

$$\sigma_{\phi,q} = \frac{2}{2^b\sqrt{3N}}$$

Source instability error (frequency and irradiance):

$$\Delta\phi_{Freq} = 2\pi\frac{d}{c}\Delta\upsilon$$

$$\sigma_{\phi,I} = \frac{1}{SNR\sqrt{N}}$$

Fringe of equal chromatic order (FECO):

$$I_t = \frac{I_{inc}}{1 + F\sin(\phi/2)}$$

Equation Summary (cont'd)

$$\phi = \frac{2\pi}{\lambda_{air}} 2nd\cos(\theta) + 2\phi_r$$

$$F = \frac{4\rho}{(1-\rho)^2}$$

$$d_2 - d_1 = \frac{\lambda_{1,m+1}}{\lambda_{1,m} - \lambda_{1,m+1}} \left(\frac{\lambda_{2,m} - \lambda_{1,m}}{2} \right)$$

Window, double pass transmission test:

$$OPD_{Measured} = 2(n-1)\delta t$$

$$OPD_{Wedge} = 2(n-1)\alpha D$$

$$\alpha = \frac{\lambda}{2(n-1)S}$$

Window wedge, two surface reflection test:

$$\alpha = \frac{\lambda}{2nS}$$

Window wedge, tilt difference method:

$$\alpha = \frac{\delta\beta}{2(n-1)} \qquad \beta_{x1} = \frac{\lambda}{S_{x1}} = \text{Tilt}$$

$$S_{x1} = \frac{S_1}{\sin(\theta)} \qquad S_{y2} = \frac{S_2}{\cos(\theta)}$$

$$\delta\beta = \sqrt{(\beta_{x1} - \beta_{x2})^2 + (\beta_{y1} - \beta_{y2})^2}$$

Prism angle error, single pass:

$$\varepsilon = \frac{\theta}{2n} \qquad \varepsilon = \frac{\delta\beta}{4n} = \frac{|\lambda/S_{y1} - \lambda/S_{y2}|}{4n}$$

Prism angle error, double pass:

$$\varepsilon = \frac{\beta_y}{4n}$$

Prism, OPD double pass:

$$OPD_{Measured} = 2(n-1)\delta t$$

Corner cubes, angle errors:

$$\varepsilon = \frac{\delta\beta}{3.266n} \qquad \varepsilon = \frac{\beta}{3.266n}$$

Equation Summary (cont'd)

Diffraction grating testing:

$$\frac{x}{\Delta x} + \frac{\delta(x,y)}{\Delta x} = m \quad \frac{x\sin(\theta)}{\lambda} + \frac{W(x,y)}{\lambda} = m$$

$$\frac{W(x,y)}{\lambda} = \frac{\delta(x,y)}{\Delta x}$$

Phase shifting with a diffraction grating:

$$\phi_{Shift} = \frac{2\pi}{\Lambda}\Delta x$$

Radius of curvature against flat, curved reference:

$$R = \frac{r_m^2}{\lambda\left(m + \frac{1}{2}\right)} \quad \Delta R = \frac{4m\lambda R^2}{d^2}$$

F-number constraints, surface testing:

$$f/\#_{Beam} = \frac{f_{Div}}{D_{Div}} \approx \frac{1}{2NA}$$

$$f/\#_{Surface} = \frac{f_S}{D_S} = \frac{R}{2D_S}$$

$$f/\#_{Beam} \le f/\#_{Surface}$$

Cylindrical optics testing:

$$r_m = \sqrt{mf\lambda} \rightarrow y_m = \sqrt{mf\lambda}$$

Absolute spherical surface testing:

$$W_{surf} = \frac{1}{2}(W_{0^\circ} + \overline{W}_{180^\circ} - W_{focus} - \overline{W}_{focus})$$

Absolute surface roughness testing:

$$\sigma_{test,N} = \frac{1}{\sqrt{N}}\sigma_{test}; \quad \sigma_{meas,N} \cong \sigma_{ref,N}; \quad \sigma_{test} = \sqrt{\sigma_{meas}^2 - \sigma_{ref,N}^2}$$

Fringe visibility, function of roughness, wavelength:

$$V = \exp\left(\frac{-8\pi^2\sigma^2}{\lambda^2}\right)$$

Equation Summary (cont'd)

Surface sag, conics and aspheres:

$$s(r) = \frac{Cr^2}{1+\sqrt{1-(1+\kappa)C^2r^2}}; \quad C = \frac{1}{R}; \quad r^2 = x^2 + y^2$$

$$s(r) = \frac{Cr^2}{1+\sqrt{1-(1+\kappa)C^2r^2}} + A_4s^4 + A_6s^6 + A_8s^8 + \cdots$$

Hindle sphere diameter:

$$D_{HS} = \frac{D(m+1)}{mo_r + 1}$$

Diffraction efficiency:

$$\eta_{Amplitude} = 0.25 \cdot \mathrm{sinc}^2(m/2) \quad \eta_{Phase} = \mathrm{sinc}^2(m/2)$$

Nyquist, cutoff frequency, G-factor:

$$\xi_{Ny} = \frac{1}{2x_s} \quad \xi_c = \frac{1}{x_s} = 2\xi_{Ny}$$

$$G = \frac{x_w}{x_s} \quad \xi_{c,sa} = \frac{1}{Gx_s}$$

Two-wavelength interferometry:

$$\lambda_{eq} = \frac{\lambda_1\lambda_2}{|\lambda_1 - \lambda_2|} \quad OPD = 2hn = \frac{\lambda_{eq}}{2\pi n}(\Delta\phi_{\lambda 1} - \Delta\phi_{\lambda 2})$$

Foucault (knife-edge) test for spherical aberration:

$$\rho = \sqrt{x^2 + y^2} = \sqrt{\frac{-\varepsilon_z r_p^2}{4R^2 W_{040}}}$$

Bibliography

Ai, C. and J.C. Wyant, "Effect of spurious reflection on phase shift interferometry," *Appl. Opt.* 27, 3039–3045 (1988).

Born, M. and E. Wolf, *Principles of Optics*, 7th Ed., Cambridge University Press, New York, 1999.

Brophy, C.P., "Effect of intensity error correlation on the computed phase of phase-shifting interferometry," *J. Opt. Soc. Am. A* 7, 537–541 (1990).

Bruning, J.H., Dr. R. Herriott, J.E. Gallagher, D.P. Rosenfeld, A.D. White, and D.J. Brangaccio, "Digital wavefront measuring interferometer for testing optical surfaces and lenses," *Appl. Opt.* 13, 2693–2703 (1974).

Caber, P.J., "Interferometric profiler for rough surfaces," *Appl. Opt.* 32, 3438–3441 (1993).

Cheng, Y. and J.C. Wyant, "Phase shifter calibration in phase-shifting interferometry," *Appl. Opt.* 24, 3049–3052 (1985).

Crane, R., "New developments in interferometry," *Appl. Opt.* 8, 538 (1969).

Creath, K. and J.C. Wyant, Ch. 16, "Moiré and fringe projection techniques," in D. Malacara, *Optical Shop Testing*, 2nd Ed., Wiley, New York, 1992.

Creath, K., "Phase-measurement interferometry techniques," in *Progress in Optics. Vol. XXVI*, pp. 349–393, E. Wolf, Ed., Elsevier Science Publishers, Amsterdam, 1988.

de Groot, P.J. and L.L. Deck, "Numerical simulation of vibration in phase-shifting interferometry," *Appl. Opt.* 35, 2172–2178 (1996).

Francon, M. and S. Mallick, *Polarization Interferometers: Applications in Microscopy and Macroscopy*, Wiley, New York, 1971.

Bibliography

Gappinger, R.O. and J.E. Greivenkamp, "Non-null interferometer for measurement of aspheric transmited wavefronts," *Proc. SPIE* 5180, 301 (2004).

Gaskill, J., *Linear Systems, Fourier Transforms, and Optics*, Wiley, New York, 1978.

Ghiglia, D.C. and M.D. Pritt, *Two-Dimensional Phase Unwrapping: Theory, Algorithms and Software*, Wiley, New York, 1998.

Greivenkamp, J.E. and J.H. Bruning. "Phase shifting interferometers," in D. Malacara, *Optical Shop Testing*, 2nd Ed., Wiley, New York, 1992.

Greivenkamp, J.E., "Sub-Nyquist interferometry," *Appl. Opt.* 26, 5245–5258 (1987).

Hariharan, P. and D. Malacara, Ed., *Selected Papers on Interference, Interferometry, and Interferometric Metrology*, SPIE Press, Bellingham, WA, 1995.

Hariharan, P., Ed., *Selected Papers on Interferometry*, SPIE Press, Bellingham, WA, 1991.

Hariharan, P., *Optical Interferometry*, 2nd Ed., Elsevier, San Diego, CA, 2003.

Jahanmir, J. and J.C. Wyant, "Comparison of surface roughness measured with an optical profiler and a scanning probe microscope," *Proc. SPIE* 1720, 111–118 (1992).

Johnson, B.K., *Optics and Optical Instruments*, Dover, New York, 1976.

Kingslake, R., B. Thompson, R.R. Shannon, and J.C. Wyant, Ed., *Applied Optics and Optical Engineering*, Vols. 1–11, Academic Press, New York, 1965–1992.

Kino, G.S. and S. Chim, "Mirau correlation microscope," *Appl. Opt.* 29, 3775–3783 (1990).

Bibliography

Koliopoulos, C., O. Kwon, R. Shagam, and J.C. Wyant, "Infrared point-diffraction interferometer," *Opt. Lett.* 3, 118–120 (1978).

Kwon, O., J.C. Wyant, and C.R. Hayslett, "Rough surface interferometry at 10.6 μm," *Appl. Opt.* 19, 1862–1869 (1980).

Malacara, D., Ed., *Selected Papers on Optical Shop Metrology*, SPIE Press, Bellingham, WA, 1990.

Malacara, D., Ed., *Optical Shop Testing*, 2nd Ed., John Wiley & Sons, New York, 1992.

Malacara, D., Ed., *Selected Papers on Optical Testing* (CD-ROM Vol. CD03), SPIE, Bellingham, WA, 1999.

Malacara, D., M. Servín, and Z. Malacara, *Interferogram Analysis for Optical Testing*, Marcel Dekker, New York, 1998.

Medecki, H., E. Tejnil, K.A. Goldberg, and J. Bokor, "Phase-shifting point diffraction interferometer," *Opt. Lett.* 21, 1526–1528 (1996).

Millerd, J. et al., "Modern approaches in phase measuring metrology," *Proc. SPIE* 5856, 14–22 (2005).

North-Morris, M.B., J. VanDelden, and J.C. Wyant, "Phase-shifting birefringent scatterplate interferometer," *Appl. Opt.* 41, 668–677 (2002).

Rimmer, M.P. and J.C. Wyant, "Evaluation of large aberrations using a lateral-shear itnerferometer having variable shear," *Appl. Opt.* 14, 142–150 (1975).

Ronchi, V., "Forty years of history of a grating interferometer," *Appl. Opt.* 3, 437–451 (1964).

Rubin, L.F. and J.C. Wyant, "Energy distribution in a scatterplate interferometer," *J. Opt. Soc. Am.* 69, 1305–1308 (1979).

Bibliography

Schwider, J., R. Burow, K.E. Elssner, J. Grzanna, R. Spolaczyk, and K. Merkel, "Digital wave-front measuring interferometry: some systematic error sources," *Appl. Opt.* 22, 3421–3432 (1983).

Shack, R.V. and G.W. Hopkins, "The Shack interferometer," *Opt. Eng.* 18, 226–228 (1979).

Shagam, R.N. and J.C. Wyant, "Optical frequency shifter for heterodyne interferometer using multiple rotating polarization retarders," *App. Opt.* 17, 3034–3035 (1978).

Smith, W.J., *Modern Optical Engineering*, McGraw-Hill, New York, 2000.

Su, D. and L. Shyu, "Phase shifting scatter plate interferometer using a polarization technique," *J. Mod. Opt.* 38, 951 (1991).

Thomas, D. and J.C. Wyant, "Determination of the dihedral angle errors of a corner cube from its Twyman-Green interferogram," *J. Opt. Soc. Am.* 67, 467–472 (1977).

Wyant, J.C. and K. Creath, "Two-wavelength phase-shifting interferometer and method," U.S. Patent 4,832,489 (1989).

Wyant, J.C. and K. Creath, Ch.1 in *Applied Optics and Optical Engineering*, Academic Press, New York, 1992.

Wyant, J.C. and V.P. Bennett, "Using computer generated holograms to test aspheric wavefronts," *Appl. Opt.* 11, 2833–2839 (1972).

Wyant, J.C., "Testing aspherics using two-wavelength holography," *Appl. Opt.* 10, 2113–2118 (1971).

Wyant, J.C., P.K. O'Neill, and A.J. MacGovern, "Interferometric method of measuring plotter distortion," *Appl. Opt.* 13, 1549–1551 (1974).

Index

Index (cont'd)

Index (cont'd)

Index (cont'd)

Eric P. Goodwin is a graduate student of the College of Optical Sciences at the University of Arizona. He graduated with his B.S. degree in optical engineering in 2002 and earned his M.S. degree in 2004, both from the University of Arizona.

While an undergraduate, Goodwin worked on gamma-ray detection for Professors H. Bradford Barber and Harrison H. Barrett. He also had internships at Lawrence Livermore National Labs (LLNL) and Nortel Networks. Goodwin's dissertation work is being done under Professor John E. Greivenkamp, where he has used multiple interferometers for various optical testing applications.

James C. Wyant received a B.S. in physics in 1965 from Case Western Reserve University and M.S. and Ph.D. in optics from the University of Rochester in 1967 and 1968. He was an optical engineer with the Itek Corporation from 1968 to 1974, when he joined the faculty of the Optical Sciences Center at the University of Arizona where he was an assistant professor 1974–1976, associate professor 1976–1979, and professor 1979–date. In 1999 he became the director of the Optical Sciences Center and in 2005 he became the first dean of the College of Optical Sciences. He was a founder of the WYKO Corporation and served as its president and board chairman from 1984 to 1997. He was a founder of 4D Technology and currently serves as its board chairman and he was a founder of DMetrix and is currently a board member.

Wyant is a fellow of the Optical Society of America (OSA), the International Optical Engineering Society (SPIE), and the Optical Society of India. Wyant was the 1986 president of SPIE and he has been an elected member of the OSA Board of Directors and Executive Committee. Wyant is currently editor-in-chief of *Applied Optics*.

Wyant has received several awards including the OSA Joseph Fraunhofer Award, 1992; SPIE Gold Medal, 2003; and the SPIE Technology Achievement Award, 1988.

Wyant's research interests include interferometry, holography, and optical testing and he has been the major advisor of 31 graduated Ph.D. students and 24 M.S. students.